These safety symbols are used in laboratory and field investigations in this bo[ok]
ing of each symbol and refer to this page often. *Remember to wash your han[ds]*

PROTECTIVE EQUIPMENT Do not begin any lab without the prope[r...]

 GOGGLES Proper eye protection must be worn when performing or observing science activities which involve items or conditions as listed below.

 APRON Wear an approved apron when using substances that could stain, wet, or destroy cloth.

 SOAP Wash hands with soap and water before removing goggles and after all lab activities.

 GLOVES Wear gloves w[hen] working with biological materials, chemicals, animals, or materials that can stain or irritate hands.

LABORATORY HAZARDS

Symbols	Potential Hazards	Precaution	Response
DISPOSAL	contamination of classroom or environment due to improper disposal of materials such as chemicals and live specimens	• DO NOT dispose of hazardous materials in the sink or trash can. • Dispose of wastes as directed by your teacher.	• If hazardous materials are disposed of improperly, notify your teacher immediately.
EXTREME TEMPERATURE	skin burns due to extremely hot or cold materials such as hot glass, liquids, or metals; liquid nitrogen; dry ice	• Use proper protective equipment, such as hot mitts and/or tongs, when handling objects with extreme temperatures.	• If injury occurs, notify your teacher immediately.
SHARP OBJECTS	punctures or cuts from sharp objects such as razor blades, pins, scalpels, and broken glass	• Handle glassware carefully to avoid breakage. • Walk with sharp objects pointed downward, away from you and others.	• If broken glass or injury occurs, notify your teacher immediately.
ELECTRICAL	electric shock or skin burn due to improper grounding, short circuits, liquid spills, or exposed wires	• Check condition of wires and apparatus for fraying or uninsulated wires, and broken or cracked equipment. • Use only GFCI-protected outlets	• DO NOT attempt to fix electrical problems. Notify your teacher immediately.
CHEMICAL	skin irritation or burns, breathing difficulty, and/or poisoning due to touching, swallowing, or inhalation of chemicals such as acids, bases, bleach, metal compounds, iodine, poinsettias, pollen, ammonia, acetone, nail polish remover, heated chemicals, mothballs, and any other chemicals labeled or known to be dangerous	• Wear proper protective equipment such as goggles, apron, and gloves when using chemicals. • Ensure proper room ventilation or use a fume hood when using materials that produce fumes. • NEVER smell fumes directly. • NEVER taste or eat any material in the laboratory.	• If contact occurs, immediately flush affected area with water and notify your teacher. • If a spill occurs, leave the area immediately and notify your teacher.
FLAMMABLE	unexpected fire due to liquids or gases that ignite easily such as rubbing alcohol	• Avoid open flames, sparks, or heat when flammable liquids are present.	• If a fire occurs, leave the area immediately and notify your teacher.
OPEN FLAME	burns or fire due to open flame from matches, Bunsen burners, or burning materials	• Tie back loose hair and clothing. • Keep flame away from all materials. • Follow teacher instructions when lighting and extinguishing flames. • Use proper protection, such as hot mitts or tongs, when handling hot objects.	• If a fire occurs, leave the area immediately and notify your teacher.
ANIMAL SAFETY	injury to or from laboratory animals	• Wear proper protective equipment such as gloves, apron, and goggles when working with animals. • Wash hands after handling animals.	• If injury occurs, notify your teacher immediately.
BIOLOGICAL	infection or adverse reaction due to contact with organisms such as bacteria, fungi, and biological materials such as blood, animal or plant materials	• Wear proper protective equipment such as gloves, goggles, and apron when working with biological materials. • Avoid skin contact with an organism or any part of the organism. • Wash hands after handling organisms.	• If contact occurs, wash the affected area and notify your teacher immediately.
FUME	breathing difficulties from inhalation of fumes from substances such as ammonia, acetone, nail polish remover, heated chemicals, and mothballs	• Wear goggles, apron, and gloves. • Ensure proper room ventilation or use a fume hood when using substances that produce fumes. • NEVER smell fumes directly.	• If a spill occurs, leave area and notify your teacher immediately.
IRRITANT	irritation of skin, mucous membranes, or respiratory tract due to materials such as acids, bases, bleach, pollen, mothballs, steel wool, and potassium permanganate	• Wear goggles, apron, and gloves. • Wear a dust mask to protect against fine particles.	• If skin contact occurs, immediately flush the affected area with water and notify your teacher.
RADIOACTIVE	excessive exposure from alpha, beta, and gamma particles	• Remove gloves and wash hands with soap and water before removing remainder of protective equipment.	• If cracks or holes are found in the container, notify your teacher immediately.

Your online portal to everything you need

connectED.mcgraw-hill.com

Look for these icons to access exciting digital resources

- Video
- Audio
- Review
- Inquiry
- WebQuest
- Assessment
- Concepts in Motion

McGraw Hill Education

LIFE: STRUCTURE AND FUNCTION

iSCIENCE

Glencoe

Snow Leopard, *Uncia uncia*
The snow leopard lives in central Asia at altitudes of 3,000 m–5,500 m. Its thick fur and broad, furry feet are two of its adaptations that make it well suited to a snowy environment. Snow leopards cannot roar but can hiss, growl, and make other sounds.

The McGraw·Hill Companies

 Education

Copyright © 2012 The McGraw-Hill Companies, Inc. All rights reserved. No part of this publication may be reproduced or distributed in any form or by any means, or stored in a database or retrieval system, without the prior written consent of The McGraw-Hill Companies, Inc., including, but not limited to, network storage or transmission, or broadcast for distance learning.

Send all inquiries to:
McGraw-Hill Education
8787 Orion Place
Columbus, OH 43240-4027

ISBN: 978-0-07-888013-1
MHID: 0-07-888013-0

Printed in the United States of America.

5 6 7 8 9 10 11 DOW 15 14

Authors and Contributors

Authors

American Museum of Natural History
New York, NY

Michelle Anderson, MS
Lecturer
The Ohio State University
Columbus, OH

Juli Berwald, PhD
Science Writer
Austin, TX

John F. Bolzan, PhD
Science Writer
Columbus, OH

Rachel Clark, MS
Science Writer
Moscow, ID

Patricia Craig, MS
Science Writer
Bozeman, MT

Randall Frost, PhD
Science Writer
Pleasanton, CA

Lisa S. Gardiner, PhD
Science Writer
Denver, CO

Jennifer Gonya, PhD
The Ohio State University
Columbus, OH

Mary Ann Grobbel, MD
Science Writer
Grand Rapids, MI

Whitney Crispen Hagins, MA, MAT
Biology Teacher
Lexington High School
Lexington, MA

Carole Holmberg, BS
Planetarium Director
Calusa Nature Center and Planetarium, Inc.
Fort Myers, FL

Tina C. Hopper
Science Writer
Rockwall, TX

Jonathan D. W. Kahl, PhD
Professor of Atmospheric Science
University of Wisconsin-Milwaukee
Milwaukee, WI

Nanette Kalis
Science Writer
Athens, OH

S. Page Keeley, MEd
Maine Mathematics and Science Alliance
Augusta, ME

Cindy Klevickis, PhD
Professor of Integrated Science and Technology
James Madison University
Harrisonburg, VA

Kimberly Fekany Lee, PhD
Science Writer
La Grange, IL

Michael Manga, PhD
Professor
University of California, Berkeley
Berkeley, CA

Devi Ried Mathieu
Science Writer
Sebastopol, CA

Elizabeth A. Nagy-Shadman, PhD
Geology Professor
Pasadena City College
Pasadena, CA

William D. Rogers, DA
Professor of Biology
Ball State University
Muncie, IN

Donna L. Ross, PhD
Associate Professor
San Diego State University
San Diego, CA

Marion B. Sewer, PhD
Assistant Professor
School of Biology
Georgia Institute of Technology
Atlanta, GA

Julia Meyer Sheets, PhD
Lecturer
School of Earth Sciences
The Ohio State University
Columbus, OH

Michael J. Singer, PhD
Professor of Soil Science
Department of Land, Air and Water Resources
University of California
Davis, CA

Karen S. Sottosanti, MA
Science Writer
Pickerington, Ohio

Paul K. Strode, PhD
I.B. Biology Teacher
Fairview High School
Boulder, CO

Jan M. Vermilye, PhD
Research Geologist
Seismo-Tectonic Reservoir Monitoring (STRM)
Boulder, CO

Judith A. Yero, MA
Director
Teacher's Mind Resources
Hamilton, MT

Dinah Zike, MEd
Author, Consultant,
Inventor of Foldables
Dinah Zike Academy;
Dinah-Might Adventures, LP
San Antonio, TX

Margaret Zorn, MS
Science Writer
Yorktown, VA

Consulting Authors

Alton L. Biggs
Biggs Educational Consulting
Commerce, TX

Ralph M. Feather, Jr., PhD
Assistant Professor
Department of Educational
Studies and Secondary
Education
Bloomsburg University
Bloomsburg, PA

Douglas Fisher, PhD
Professor of Teacher Education
San Diego State University
San Diego, CA

Edward P. Ortleb
Science/Safety Consultant
St. Louis, MO

Series Consultants

Science

Solomon Bililign, PhD
Professor
Department of Physics
North Carolina Agricultural
and Technical State University
Greensboro, NC

John Choinski
Professor
Department of Biology
University of Central Arkansas
Conway, AR

Anastasia Chopelas, PhD
Research Professor
Department of Earth and
Space Sciences
UCLA
Los Angeles, CA

David T. Crowther, PhD
Professor of Science Education
University of Nevada, Reno
Reno, NV

A. John Gatz
Professor of Zoology
Ohio Wesleyan University
Delaware, OH

Sarah Gille, PhD
Professor
University of California
San Diego
La Jolla, CA

David G. Haase, PhD
Professor of Physics
North Carolina State
University
Raleigh, NC

Janet S. Herman, PhD
Professor
Department of Environmental
Sciences
University of Virginia
Charlottesville, VA

David T. Ho, PhD
Associate Professor
Department of Oceanography
University of Hawaii
Honolulu, HI

Ruth Howes, PhD
Professor of Physics
Marquette University
Milwaukee, WI

Jose Miguel Hurtado, Jr., PhD
Associate Professor
Department of Geological
Sciences
University of Texas at El Paso
El Paso, TX

Monika Kress, PhD
Assistant Professor
San Jose State University
San Jose, CA

Mark E. Lee, PhD
Associate Chair & Assistant
Professor
Department of Biology
Spelman College
Atlanta, GA

Linda Lundgren
Science writer
Lakewood, CO

Series Consultants, continued

Keith O. Mann, PhD
Ohio Wesleyan University
Delaware, OH

Charles W. McLaughlin, PhD
Adjunct Professor of Chemistry
Montana State University
Bozeman, MT

Katharina Pahnke, PhD
Research Professor
Department of Geology and Geophysics
University of Hawaii
Honolulu, HI

Jesús Pando, PhD
Associate Professor
DePaul University
Chicago, IL

Hay-Oak Park, PhD
Associate Professor
Department of Molecular Genetics
Ohio State University
Columbus, OH

David A. Rubin, PhD
Associate Professor of Physiology
School of Biological Sciences
Illinois State University
Normal, IL

Toni D. Sauncy
Assistant Professor of Physics
Department of Physics
Angelo State University
San Angelo, TX

Malathi Srivatsan, PhD
Associate Professor of Neurobiology
College of Sciences and Mathematics
Arkansas State University
Jonesboro, AR

Cheryl Wistrom, PhD
Associate Professor of Chemistry
Saint Joseph's College
Rensselaer, IN

Reading

ReLeah Cossett Lent
Author/Educational Consultant
Blue Ridge, GA

Math

Vik Hovsepian
Professor of Mathematics
Rio Hondo College
Whittier, CA

Series Reviewers

Thad Boggs
Mandarin High School
Jacksonville, FL

Catherine Butcher
Webster Junior High School
Minden, LA

Erin Darichuk
West Frederick Middle School
Frederick, MD

Joanne Hedrick Davis
Murphy High School
Murphy, NC

Anthony J. DiSipio, Jr.
Octorara Middle School
Atglen, PA

Adrienne Elder
Tulsa Public Schools
Tulsa, OK

Series Reviewers, continued

Carolyn Elliott
Iredell-Statesville Schools
Statesville, NC

Christine M. Jacobs
Ranger Middle School
Murphy, NC

Jason O. L. Johnson
Thurmont Middle School
Thurmont, MD

Felecia Joiner
Stony Point Ninth Grade Center
Round Rock, TX

Joseph L. Kowalski, MS
Lamar Academy
McAllen, TX

Brian McClain
Amos P. Godby High School
Tallahassee, FL

Von W. Mosser
Thurmont Middle School
Thurmont, MD

Ashlea Peterson
Heritage Intermediate Grade Center
Coweta, OK

Nicole Lenihan Rhoades
Walkersville Middle School
Walkersvillle, MD

Maria A. Rozenberg
Indian Ridge Middle School
Davie, FL

Barb Seymour
Westridge Middle School
Overland Park, KS

Ginger Shirley
Our Lady of Providence Junior-Senior High School
Clarksville, IN

Curtis Smith
Elmwood Middle School
Rogers, AR

Sheila Smith
Jackson Public School
Jackson, MS

Sabra Soileau
Moss Bluff Middle School
Lake Charles, LA

Tony Spoores
Switzerland County Middle School
Vevay, IN

Nancy A. Stearns
Switzerland County Middle School
Vevay, IN

Kari Vogel
Princeton Middle School
Princeton, MN

Alison Welch
Wm. D. Slider Middle School
El Paso, TX

Linda Workman
Parkway Northeast Middle School
Creve Coeur, MO

Teacher Advisory Board

The Teacher Advisory Board gave the authors, editorial staff, and design team feedback on the content and design of the Student Edition. They provided valuable input in the development of *Glencoe ⓘScience*.

Frances J. Baldridge
Department Chair
Ferguson Middle School
Beavercreek, OH

Jane E. M. Buckingham
Teacher
Crispus Attucks Medical
Magnet High School
Indianapolis, IN

Elizabeth Falls
Teacher
Blalack Middle School
Carrollton, TX

Nelson Farrier
Teacher
Hamlin Middle School
Springfield, OR

Michelle R. Foster
Department Chair
Wayland Union
Middle School
Wayland, MI

Rebecca Goodell
Teacher
Reedy Creek Middle School
Cary, NC

Mary Gromko
Science Supervisor K–12
Colorado Springs District 11
Colorado Springs, CO

Randy Mousley
Department Chair
Dean Ray Stucky
Middle School
Wichita, KS

David Rodriguez
Teacher
Swift Creek Middle School
Tallahassee, FL

Derek Shook
Teacher
Floyd Middle Magnet School
Montgomery, AL

Karen Stratton
Science Coordinator
Lexington School District One
Lexington, SC

Stephanie Wood
Science Curriculum Specialist,
K–12
Granite School District
Salt Lake City, UT

Online Guide

connectED.mcgraw-hill.com

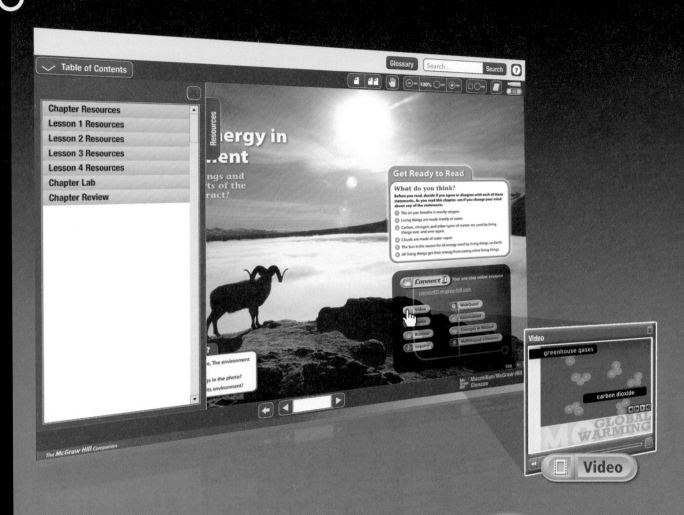

Your Digital Science Portal

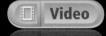

See the science in real life through these exciting videos.

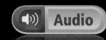

Click the link and you can listen to the text while you follow along.

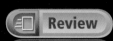

Try these interactive tools to help you review the lesson concepts.

Explore concepts through hands-on and virtual labs.

These web-based challenges relate the concepts you're learning about to the latest news and research.

Digital and Print Solutions

The icons in your online student edition link you to interactive learning opportunities. Browse your online student book to find more.

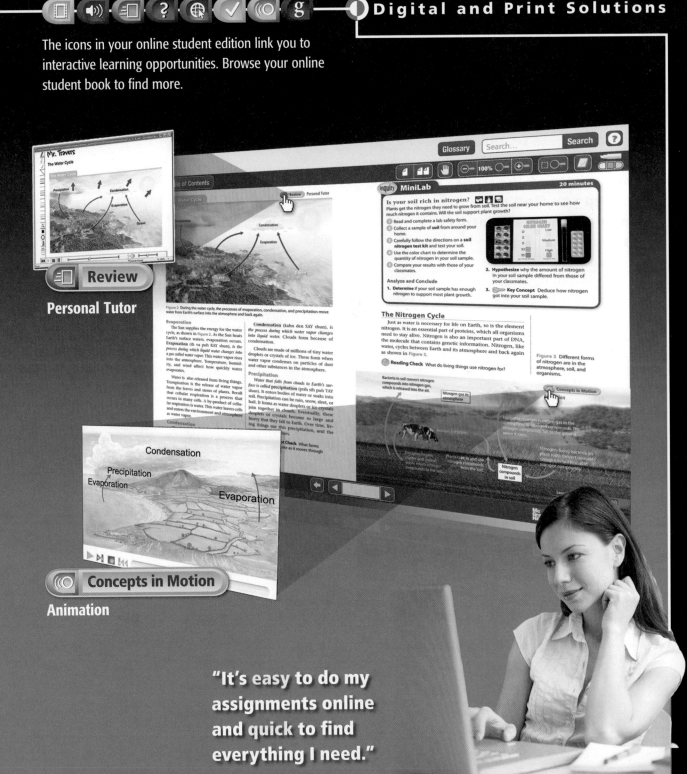

Personal Tutor

Animation

"It's easy to do my assignments online and quick to find everything I need."

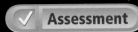 **Assessment**

Check how well you understand the concepts with online quizzes and practice questions.

 Concepts in Motion

The textbook comes alive with animated explanations of important concepts.

 Multilingual eGlossary

Read key vocabulary in 13 languages.

Treasure Hunt

Your science book has many features that will aid you in your learning. Some of these features are listed below. You can use the activity at the right to help you find these and other special features in the book.

- **THE BIG IDEA** can be found at the start of each chapter.
- The Reading Guide at the start of each lesson lists 🔑 **Key Concepts**, vocabulary terms, and online supplements to the content.
- **ConnectED** icons direct you to online resources such as animations, personal tutors, math practices, and quizzes.
- **Inquiry** Labs and Skill Practices are in each chapter.
- Your **FOLDABLES** help organize your notes.

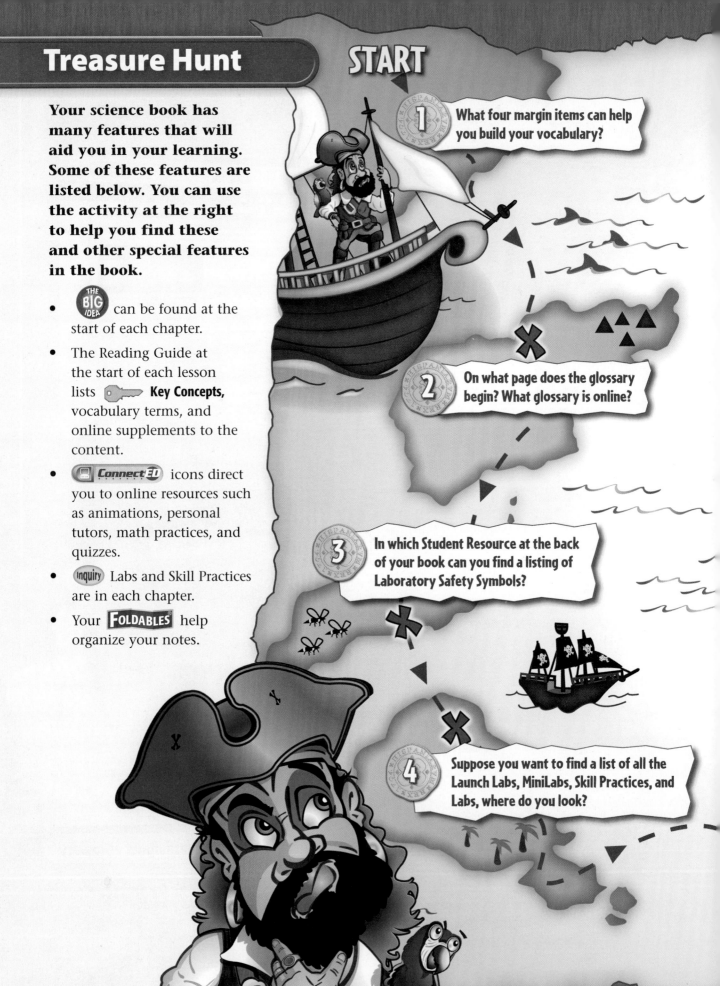

START

1. What four margin items can help you build your vocabulary?

2. On what page does the glossary begin? What glossary is online?

3. In which Student Resource at the back of your book can you find a listing of Laboratory Safety Symbols?

4. Suppose you want to find a list of all the Launch Labs, MiniLabs, Skill Practices, and Labs, where do you look?

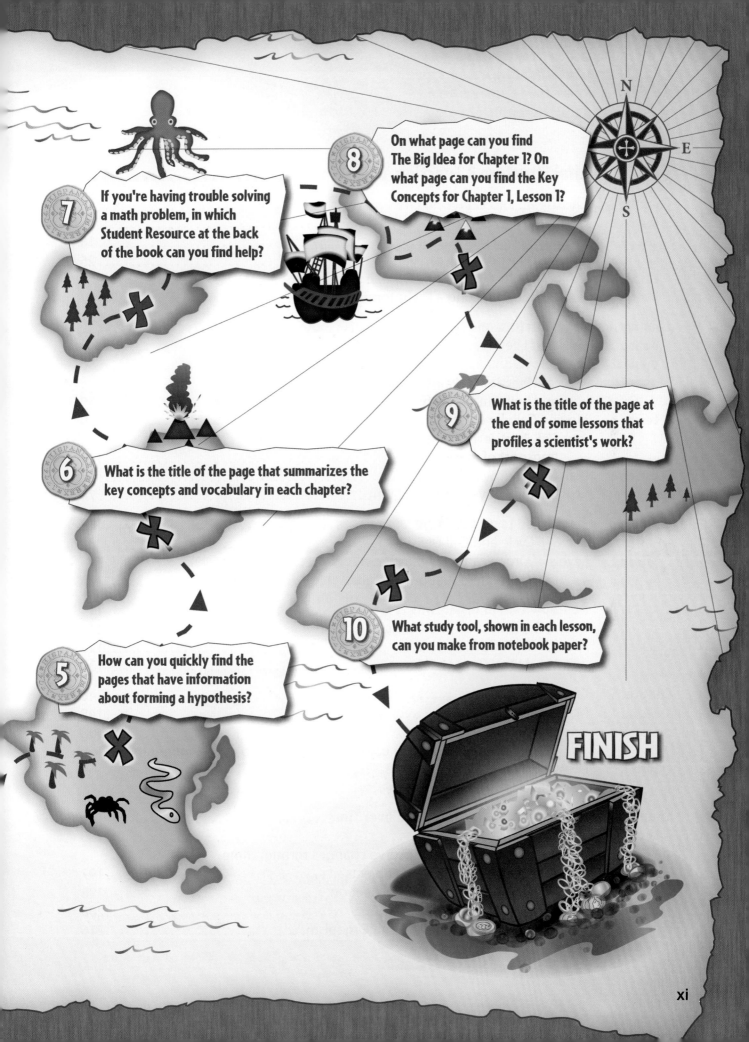

Table of Contents

Unit 1 Life: Structure & Function..2

Chapter 1 **Classifying and Exploring Life** 6
- **Lesson 1** Characteristics of Life 8
- **Lesson 2** Classifying Organisms...18
  - **Skill Practice** How can you identify a beetle? 25
- **Lesson 3** Exploring Life 26
  - **Lab** Constructing a Dichotomous Key.......................................32

Chapter 2 **Cell Structure and Function**..40
- **Lesson 1** Cells and Life 42
- **Lesson 2** The Cell..50
  - **Skill Practice** How are plant cells and animal cells similar and how are they different?...59
- **Lesson 3** Moving Cellular Material 60
  - **Skill Practice** How does an object's size affect the transport of materials?...67
- **Lesson 4** Cells and Energy 68
  - **Lab** Photosynthesis and Light..74

Chapter 3 **From a Cell to an Organism** 82
- **Lesson 1** The Cell Cycle and Cell Division.....................................84
- **Lesson 2** Levels of Organization 96
  - **Lab** Cell Differentiation...106

Chapter 4 **Reproduction of Organisms** 114
- **Lesson 1** Sexual Reproduction and Meiosis 116
- **Lesson 2** Asexual Reproduction...128
  - **Lab** Mitosis and Meiosis 138

Chapter 5 **Genetics**...146
- **Lesson 1** Mendel and His Peas 148
- **Lesson 2** Understanding Inheritance...158
  - **Skill Practice** How can you use Punnett squares to model inheritance?..168
- **Lesson 3** DNA and Genetics 169
  - **Lab** Gummy Bear Genetics 178

Chapter 6 **The Environment and Change Over Time**.........................186
- **Lesson 1** Fossil Evidence of Evolution 188
  - **Skill Practice** Can you observe changes through time in collections of everyday objects?..197
- **Lesson 2** Theory of Evolution by Natural Selection 198
- **Lesson 3** Biological Evidence of Evolution...................................208
  - **Lab** Model Adaptations in an Organism....................................216

Table of Contents

Student Resources

Science Skill Handbook SR-2
  Scientific Methods SR-2
  Safety Symbols SR-11
  Safety in the Science Laboratory SR-12

Math Skill Handbook SR-14
  Math Review SR-14
  Science Applications SR-24

Reference Handbook SR-29
  Use and Care of a Microscope SR-29
  Diversity of Life: Classification of Living Organisms SR-30
  Periodic Table of the Elements SR-34

Glossary G-2
Index I-2
Credits C-2

Inquiry

Launch Labs

1-1	Is it alive?...	9
1-2	How do you identify similar items?..	19
1-3	Can a water drop make objects appear bigger or smaller?...................	27
2-1	What's in a cell?..	43
2-2	Why do eggs have shells?..	51
2-3	What does the cell membrane do?...	61
2-4	What do you exhale?...	69
3-1	Why isn't your cell like mine?..	85
3-2	How is a system organized?..	97
4-1	Why do offspring look different?..	117
4-2	How do yeast reproduce?...	129
5-1	What makes you unique?..	149
5-2	What is the span of your hand?..	159
5-3	How are codes used to determine traits?...................................	170
6-1	How do fossils form?..	189
6-2	Are there variations within your class?...................................	199
6-3	How is the structure of a spoon related to its function?..................	209

MiniLabs

1-1	Did you blink?..	12
1-2	How would you name an unknown organism?...................................	23
1-3	How do microscopes help us compare living things?.........................	30
2-1	How can you observe DNA?..	47
2-2	How do eukaryotic and prokaryotic cells compare?..........................	54
2-3	How is a balloon like a cell membrane?....................................	63
3-1	How does mitosis work?..	93
3-2	How do cells work together to make an organism?...........................	103
4-1	How does one cell produce four cells?.....................................	119
4-2	What parts of plants can grow?..	133
5-1	Which is the dominant trait?..	155
5-2	Can you infer genotype?...	161
5-3	How can you model DNA?..	172
6-1	How do species change over time?..	195
6-2	Who survives?...	205
6-3	How related are organisms?..	213

Inquiry

Inquiry Skill Practice

- 1-2 How can you identify a beetle? 25
- 2-2 How are plant cells and animal cells similar and how are they different? 59
- 2-3 How does an object's size affect the transport of materials? 67
- 5-2 How can you use Punnett squares to model inheritance? 168
- 6-1 Can you observe changes through time in collections of everyday objects? 197

Inquiry Labs

- 1-3 Constructing a Dichotomous Key 32
- 2-4 Photosynthesis and Light 74
- 3-2 Cell Differentiation 106
- 4-2 Mitosis and Meiosis 138
- 5-3 Gummy Bear Genetics 178
- 6-3 Model Adaptations in an Organism 216

Features

How Nature Works

- 2-1 A Very Powerful Microscope 49

Science & Society

- 3-1 DNA Fingerprinting 95
- 5-1 Pioneering the Science of Genetics 157

Careers in Science

- 1-1 The Amazing Adaptation of an Air-Breathing Catfish 17
- 4-1 The Spider Mating Dance 127
- 6-2 Peter and Rosemary Grant 207

Unit 1

LIFE: Structure & Function

"Steer clear of those mitochondria. Too much energy!"

"Passing through some lysosomes. Their enzymes digest food particles."

"Now entering cell!"

1665
Robert Hooke discovers cells while examining thin slices of cork under a microscope.

1674
Anton van Leeuwenhoek observes living cells under a microscope and names the moving organisms *animalcules*.

1831
The nucleus is given its name by Robert Brown.

1839
Theodor Schwann publishes a book suggesting that the cell is the basic unit of life.

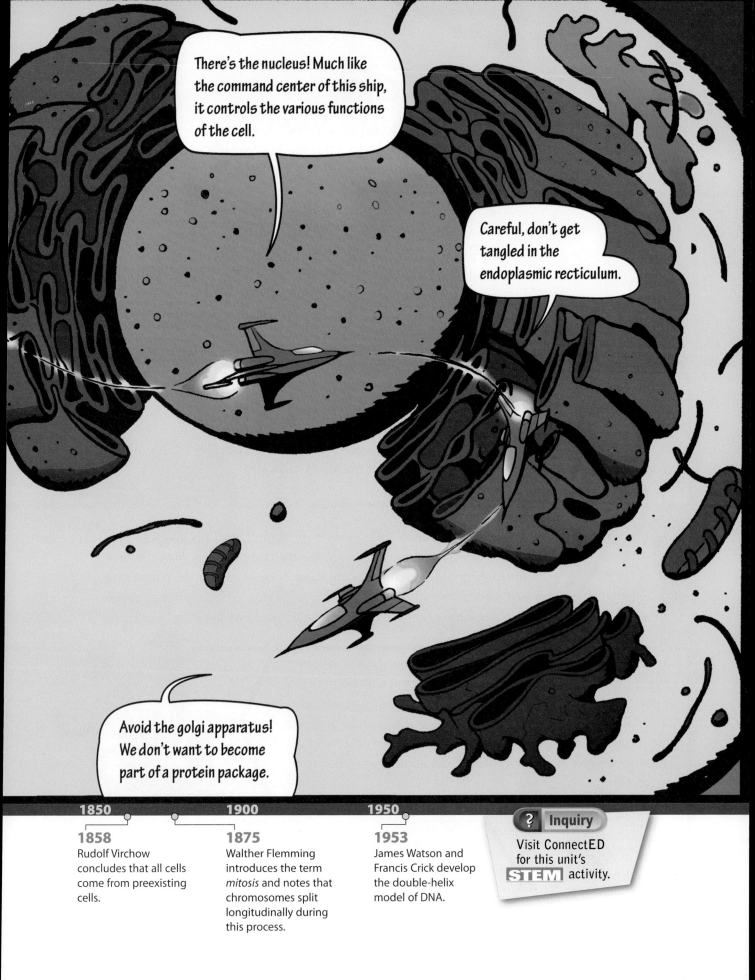

1850	1900	1950	
1858 Rudolf Virchow concludes that all cells come from preexisting cells.	**1875** Walther Flemming introduces the term *mitosis* and notes that chromosomes split longitudinally during this process.	**1953** James Watson and Francis Crick develop the double-helix model of DNA.	**Inquiry** Visit ConnectED for this unit's STEM activity.

Unit 1: Nature of SCIENCE

Models

What would you do without your heart—one of the most important muscles in your body? Worldwide, people are on donor lists, patiently waiting for heart transplants because their hearts are not working properly. Today, doctors can diagnose and treat heart problems with the help of models.

A **model** is a representation of an object, a process, an event, or a system that is similar to the physical object or idea being studied. Models can be used to study things that are too big or too small, happen too quickly or too slowly, or are too dangerous or too expensive to study directly. However, some models can replace organs or bones in the body that are not functioning properly.

A magnetic resonance image (MRI) is a type of model created by using a strong magnetic field and radio waves. MRI machines produce high-resolution images of the body from a series of images of different layers of the heart. For example, an MRI model of the heart allows cardiologists to diagnose heart disease or damage. To obtain a clear MRI, the patient must be still. Even the beating of the heart can limit the ability of an MRI to capture clear images.

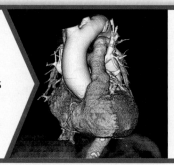

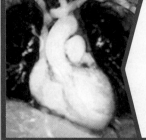

A computer tomography (CT) scan combines multiple X-ray images into a detailed 3-D visual model of structures in the body. Cardiologists use this model to diagnose a malfunctioning heart or blocked arteries. A limitation of a CT scan is that some coronary artery diseases, especially if they do not involve a buildup of calcium, may not be detected by the scan.

An artificial heart is a physical model of a human heart that can pump blood throughout the body. For a patient with heart failure, a doctor might suggest temporarily replacing the heart with an artificial model while they wait for a transplant. Because of its size, the replacement heart is suitable for about only 50 percent of the male population. And, it is stable only for about 2 years before it wears out.

A cardiologist might use a physical model of a heart to explain a diagnosis to a patient. The parts of the heart can be touched and manipulated to explain how a heart works and the location of any complications. However, this physical model does not function like a real heart, and it cannot be used to diagnose disease.

Maps as Models

One way to think of a computer model, such as an MRI or a CT scan, is as a map. A map is a model that shows how locations are arranged in space. A map can be a model of a small area, such as your street. Or, maps can be models of very large areas, such as a state, a country, or the world.

Biologists study maps to understand where different animal species live, how they interact, and how they migrate. Most animals travel in search of food, water, specific weather, or a place to mate. By placing small electronic tracking devices on migrating animals biologists can create maps of their movements, such as the map of elephant movement in **Figure 1**. These maps are models that help determine how animals survive, repeat the patterns of their life cycle, and respond to environmental changes.

Limitations of Models

It is impossible to include all the details about an object or an idea in one model. A map of elephant migration does not tell you whether the elephant is eating, sleeping, or playing with other elephants. Scientists must consider the limitations of the models they use when drawing conclusions about animal behavior.

All models have limitations. When making decisions about a patient's diagnosis and treatment, a cardiologist must be aware of the information each type of model does and does not provide. CT scans and MRIs each provide different diagnostic information. A doctor needs to know what information is needed before choosing which model to use. Scientists and doctors consider the purpose and limitations of the models they use to ensure that they draw the most accurate conclusions possible.

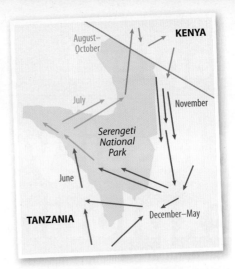

Figure 1 This map is a model of elephants' movements. The colored lines show the paths of three elephants that were equipped with tracking devices for a year.

Inquiry MiniLab   40 minutes

How can you model an elephant enclosure?

You are part of a zoo design firm hired to design a model of a new elephant enclosure that mimics a natural habitat.

1. Read and complete a lab safety form.
2. Research elephants and study the map above to understand the needs of elephants.
3. Create a detailed map of your enclosure using **colored pencils** and a **ruler**. Be sure to include the scale, labels, and a legend.
4. Trade maps with a classmate.
5. Using **salt dough** and **craft supplies**, build a physical 3-D model of the elephant enclosure.

Analyze and Conclude

1. **Describe** How did you decide on the scale for your map?
2. **Compare** What are some similarities between your map and your physical model?
3. **Contrast** What are the benefits and the limitations of your physical model?

Chapter 1

Classifying and Exploring Life

THE BIG IDEA What are living things, and how can they be classified?

Inquiry Dropped Dinner Rolls?

At first glance, you might think someone dropped dinner rolls on a pile of rocks. These objects might look like dinner rolls, but they're not.

- What do you think the objects are? Do you think they are alive?
- Why do you think they look like this?
- What are living things, and how can they be classified?

Get Ready to Read

What do you think?
Before you read, decide if you agree or disagree with each of these statements. As you read this chapter, see if you change your mind about any of the statements.

1. All living things move.
2. The Sun provides energy for almost all organisms on Earth.
3. A dichotomous key can be used to identify an unknown organism.
4. Physical similarities are the only traits used to classify organisms.
5. Most cells are too small to be seen with the unaided eye.
6. Only scientists use microscopes.

ConnectED Your one-stop online resource

connectED.mcgraw-hill.com

- Video
- Audio
- Review
- Inquiry
- WebQuest
- Assessment
- Concepts in Motion
- Multilingual eGlossary

Lesson 1

Reading Guide

Key Concepts
ESSENTIAL QUESTIONS

- What characteristics do all living things share?

Vocabulary
organism p. 9
cell p. 10
unicellular p. 10
multicellular p. 10
homeostasis p. 13

g Multilingual eGlossary

Characteristics of Life

This toy looks like a dog and can move, but it is a robot. What characteristics are missing to make it alive? Let's find out.

Inquiry Launch Lab

15 minutes

Is it alive?

Living organisms have specific characteristics. Is a rock a living organism? Is a dog? What characteristics describe something that is living?

1. Read and complete a lab safety form.
2. Place three pieces of **pasta** in the bottom of a **clear plastic cup.**
3. Add **carbonated water** to the cup until it is 2/3 full.
4. Observe the contents of the cup for 5 minutes. Record your observations in your Science Journal.

Think About This

1. Think about living things. How do you know they are alive?
2. Which characteristics of life do you think you are observing in the cup?
3.  **Key Concept** Is the pasta alive? How do you know?

Characteristics of Life

Look around your classroom and then at **Figure 1.** You might see many nonliving things, such as lights and books. Look again, and you might see many living things, such as your teacher, your classmates, and plants. What makes people and plants different from lights and books?

People and plants, like all living things, have all the characteristics of life. All living things are organized, grow and develop, reproduce, respond, maintain certain internal conditions, and use energy. Nonliving things might have some of these characteristics, but they do not have all of them. Books might be organized into chapters, and lights use energy. However, only those things that have all the characteristics of life are living. *Things that have all the characteristics of life are called* **organisms.**

 Reading Check How do living things differ from nonliving things?

Figure 1 A classroom might contain living and nonliving things.

Lesson 1
EXPLORE

FOLDABLES

Fold a sheet of paper into a half book. Label it as shown. Use it to organize your notes on the characteristics of living things.

Organization

Your home is probably organized in some way. For example, the kitchen is for cooking, and the bedrooms are for sleeping. Living things are also organized. Whether an organism is made of one **cell**—*the smallest unit of life*—or many cells, all living things have structures that have specific functions.

Living things that are made of only one cell are called **unicellular** *organisms.* Within a unicellular organism are structures with specialized functions just like a house has rooms for different activities. Some structures take in nutrients or control cell activities. Other structures enable the organism to move.

Living things that are made of two or more cells are called **multicellular** *organisms.* Some multicellular organisms only have a few cells, but others have trillions of cells. The different cells of a multicellular organism usually do not perform the same function. Instead, the cells are organized into groups that have specialized functions, such as digestion or movement.

Growth and Development

The tadpole in **Figure 2** is not a frog, but it will soon lose its tail, grow legs, and become an adult frog. This happens because the tadpole, like all organisms, will grow and develop. When organisms grow, they increase in size. A unicellular organism grows as the cell increases in size. Multicellular organisms grow as the number of their cells increases.

Figure 2 A tadpole grows in size while developing into an adult frog.

Growth and Development   Concepts in Motion Animation

✓ **Visual Check** What characteristics of life can you identify in this figure?

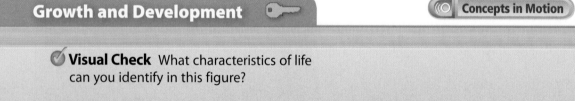

① A frog egg develops into a tadpole.

② As the tadpole grows, it develops legs.

10 Chapter 1 EXPLAIN

Changes that occur in an organism during its lifetime are called development. In multicellular organisms, development happens as cells become specialized into different cell types, such as skin cells or muscle cells. Some organisms undergo dramatic developmental changes over their lifetime, such as a tadpole developing into a frog.

 Reading Check What happens in development?

Reproduction

As organisms grow and develop, they usually are able to reproduce. Reproduction is the process by which one organism makes one or more new organisms. In order for living things to continue to exist, organisms must reproduce. Some organisms within a population might not reproduce, but others must reproduce if the species is to survive.

Organisms do not all reproduce in the same way. Some organisms, like the ones in **Figure 3,** can reproduce by dividing and become two new organisms. Other organisms have specialized cells for reproduction. Some organisms must have a mate to reproduce, but others can reproduce without a mate. The number of offspring produced varies. Humans usually produce only one or two offspring at a time. Other organisms, such as the frog in **Figure 2,** can produce hundreds of offspring at one time.

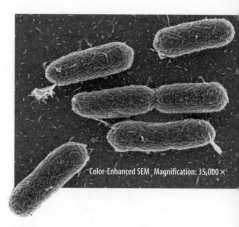

Figure 3 Some unicellular organisms, like the bacteria shown here, reproduce by dividing. The two new organisms are identical to the original organism.

3 The tadpole continues to grow as it develops into an adult frog.

4 An adult female frog can produce hundreds of eggs.

Responses to Stimuli

If someone throws a ball toward you, you might react by trying to catch it. This is because you, like all living things, respond to changes in the environment. These changes can be internal or external and are called stimuli (STIHM yuh li).

Internal Stimuli

You respond to internal stimuli (singular, stimulus) every day. If you feel hungry and then look for food, you are responding to an internal stimulus—the feeling of hunger. The feeling of thirst that causes you to find and drink water is another example of an internal stimulus.

External Stimuli

Changes in an organism's environment that affect the organism are external stimuli. Some examples of external stimuli are light and temperature.

Many plants, like the one in **Figure 4**, will grow toward light. You respond to light, too. Your skin's response to sunlight might be to darken, turn red, or freckle.

Some animals respond to changes in temperature. The response can be more or less blood flowing to the skin. For example, if the temperature increases, the diameter of an animal's blood vessels increases. This allows more blood to flow to the skin, cooling an animal.

Figure 4 The leaves and stems of plants like this one will grow toward a light source.

Inquiry MiniLab

20 minutes

Did you blink?

Like all living organisms, you respond to changes, or stimuli, in your environment. When you react to a stimulus without thinking, the response is known as a reflex. Let's see what a reflex is like.

1. Read and complete a lab safety form.
2. Sit on a chair with your hands in your lap.
3. Have your partner gently toss a **soft, foam ball** at your face five times. Your partner will warn you when he or she is going to toss the ball. Record your responses in your Science Journal.
4. Have your partner gently toss the ball at your face five times without warning you. Record your responses.
5. Switch places with your partner, and repeat steps 3 and 4.

Analyze and Conclude

1. **Compare** your responses when you were warned and when you were not warned.
2. **Decide** if any of your reactions were reflex responses, and explain your answer.
3. **Key Concept** Infer why organisms have reflex responses to some stimuli.

Homeostasis

Have you ever noticed that if you drink more water than usual, you have to go to the bathroom more often? That is because your body is working to keep your internal environment under normal conditions. *An organism's ability to maintain steady internal conditions when outside conditions change is called* **homeostasis** (hoh mee oh STAY sus).

The Importance of Homeostasis

Are there certain conditions you need to do your homework? Maybe you need a quiet room with a lot of light. Cells also need certain conditions to function properly. Maintaining certain conditions—homeostasis—ensures that cells can function. If cells cannot function normally, then an organism might become sick or even die.

Methods of Regulation

A person might not survive if his or her body temperature changes more than a few degrees from 37°C. When your outside environment becomes too hot or too cold, your body responds. It sweats, shivers, or changes the flow of blood to maintain a body temperature of 37°C.

Unicellular organisms, such as the paramecium in **Figure 5**, also have ways of regulating homeostasis. A structure called a contractile vacuole (kun TRAK tul • VA kyuh wohl) collects and pumps excess water out of the cell.

WORD ORIGIN

homeostasis
from Greek *homoios*, means "like, similar"; and *stasis*, means "standing still"

Figure 5 This paramecium lives in freshwater. Water continuously enters its cell and collects in contractile vacuoles. The vacuoles contract and expel excess water from the cell. This maintains normal water levels in the cell.

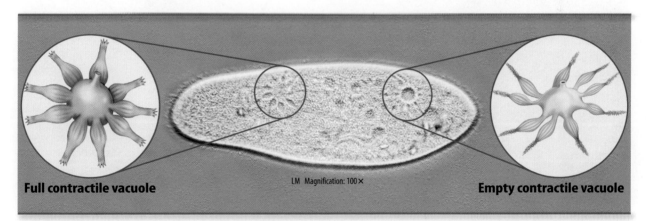

Full contractile vacuole LM Magnification: 100× **Empty contractile vacuole**

There is a limit to the amount of change that can occur within an organism. For example, you are able to survive only a few hours in water that is below 10°C. No matter what your body does, it cannot maintain steady internal conditions, or homeostasis, under these circumstances. As a result, your cells lose their ability to function.

 Reading Check Why is maintaining homeostasis important to organisms?

Energy

Everything you do requires energy. Digesting your food, sleeping, thinking, reading and all of the characteristics of life shown in **Table 1** on the next page require energy. Cells continuously use energy to transport substances, make new cells, and perform chemical reactions. Where does this energy come from?

For most organisms, this energy originally came to Earth from the Sun, as shown in **Figure 6**. For example, energy in the cactus came from the Sun. The squirrel gets energy by eating the cactus, and the coyote gets energy by eating the squirrel.

Key Concept Check What characteristics do all living things share?

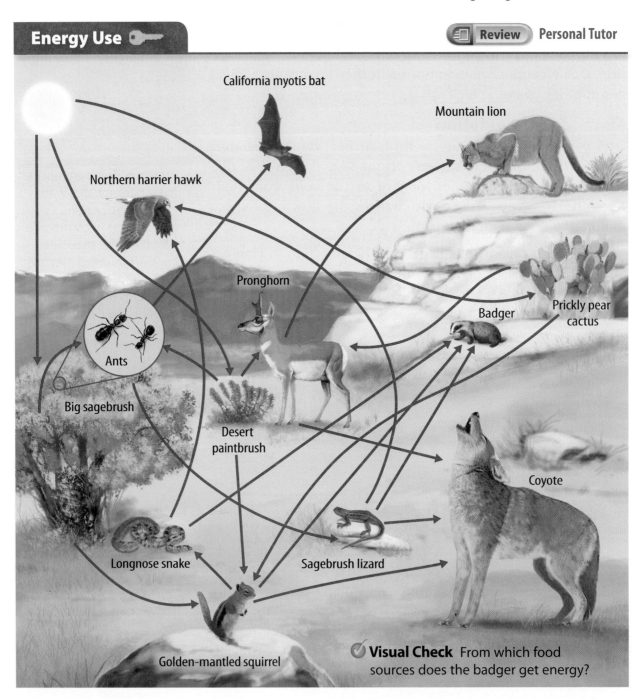

Visual Check From which food sources does the badger get energy?

Figure 6 All organisms require energy to survive. In this food web, energy passes from one organism to another and to the environment.

Table 1 Characteristics of Life

Concepts in Motion Interactive Table

Characteristic	Definition	Example
Organization	Living things have specialized structures with specialized functions. Living things with more than one cell have a greater level of organization because groups of cells function together.	
Growth and development	Living things grow by increasing cell size and/or increasing cell number. Multicellular organisms develop as cells develop specialized functions.	
Reproduction	Living things make more living things through the process of reproduction.	
Response to stimuli	Living things adjust and respond to changes in their internal and external environments.	
Homeostasis	Living things maintain stable internal conditions.	
Use of energy	Living things use energy for all the processes they perform. Living things get energy by making their own food, eating food, or absorbing food.	

Lesson 1
EXPLAIN

Lesson 1 Review

Assessment Online Quiz

Visual Summary

An organism has all the characteristics of life.

Unicellular organisms have specialized structures, much like a house has rooms for different activities.

Homeostasis enables living things to maintain a steady internal environment.

FOLDABLES

Use your lesson Foldable to review the lesson. Save your Foldable for the project at the end of the chapter.

Use Vocabulary

1. A(n) _____ is the smallest unit of life.
2. **Distinguish** between unicellular and multicellular.
3. **Define** the term *homeostasis* in your own words.

Understand Key Concepts

4. Which is NOT a characteristic of all living things?
   A. breathing   C. reproducing
   B. growing   D. using energy

5. **Compare** the processes of reproduction and growth.

6. **Choose** the characteristic of living things that you think is most important. Explain why you chose that characteristic.

7. **Critique** the following statement: A candle flame is a living thing.

Interpret Graphics

8. **Summarize** Copy and fill in the graphic organizer below to summarize the characteristics of living things.

9. **Describe** all the characteristics of life that are represented in the figure below.

Critical Thinking

10. **Suggest** how organisms would be different if they were not organized.

11. **Hypothesize** what would happen if living things could not reproduce.

What do you think NOW?

You first read the statements below at the beginning of the chapter.

1. All living things move.

2. The Sun provides energy for almost all organisms on Earth.

Did you change your mind about whether you agree or disagree with the statements? Rewrite any false statements to make them true.

CAREERS in SCIENCE

The Amazing Adaptation of an Air-Breathing Catfish

Discover how some species of armored catfish breathe air.

Have you ever thought about why animals need oxygen? All animals, including you, get their energy from food. When you breathe, the oxygen you take in is used in your cells. Chemical reactions in your cells use oxygen and change the energy in food molecules into energy that your cells can use. Mammals and many other animals get oxygen from air. Most fish get oxygen from water. Either way, after an animal takes in oxygen, red blood cells carry oxygen to cells throughout its body.

Adriana Aquino is an ichthyologist (IHK thee AH luh jihst) at the American Museum of Natural History in New York City. She discovers and classifies species of fish, such as the armored catfish in family Loricariidae from South America. It lives in freshwater rivers and pools in the Amazon. Its name comes from the bony plates covering its body. Some armored catfish can take in oxygen from water and from air!

Some armored catfish live in fast-flowing rivers. The constant movement of the water evenly distributes oxygen throughout it. The catfish can easily remove oxygen from this oxygen-rich water.

But other armored catfish live in pools of still water, where most oxygen is only at the water's surface. This makes the pools low in oxygen. To maintain a steady level of oxygen in their cells, these fish have adaptations that enable them to take in oxygen directly from air. These catfish can switch from removing oxygen from water through their gills to removing oxygen from air through the walls of their stomachs. They can only do this when they do not have much food in their stomachs. Some species can survive up to 30 hours out of water!

Meet an Ichthyologist

Aquino examines hundreds of catfish specimens. Some she collects in the field, and others come from museum collections. She compares the color, the size, and the shape of the various species. She also examines their internal and external features, such as muscles, gills, and bony plates.

Some armored catfish remove oxygen from air.

It's Your Turn

BRAINSTORM Work with a group. Choose an animal and list five physical characteristics. Brainstorm how these adaptations help the animal be successful in its habitat. Present your findings to the class.

Lesson 1 EXTEND

Lesson 2

Reading Guide

Key Concepts
ESSENTIAL QUESTIONS
- What methods are used to classify living things into groups?
- Why does every species have a scientific name?

Vocabulary
binomial nomenclature p. 21
species p. 21
genus p. 21
dichotomous key p. 22
cladogram p. 23

 Multilingual eGlossary

 Video   BrainPOP®

Classifying Organisms

Inquiry   Alike or Not?

In a band, instruments are organized into groups, such as brass and woodwinds. The instruments in a group are alike in many ways. In a similar way, living things are classified into groups. Why are living things classified?

Launch Lab

15 minutes

How do you identify similar items?

Do you separate your candies by color before you eat them? When your family does laundry, do you sort the clothes by color first? Identifying characteristics of items can enable you to place them into groups.

1. Read and complete a lab safety form.
2. Examine twelve **leaves.** Choose a characteristic that you could use to separate the leaves into two groups. Record the characteristic in your Science Journal.
3. Place the leaves into two groups, *A* and *B*, using the characteristic you chose in step 2.
4. Choose another characteristic that you could use to further divide group A. Record the characteristic, and divide the leaves.
5. Repeat step 4 with group B.

Think About This

1. What types of characteristics did other groups in class choose to separate the leaves?
2.  **Key Concept** Why would scientists need rules for separating and identifying items?

Classifying Living Things

How would you find your favorite fresh fruit or vegetable in the grocery store? You might look in the produce section, such as the one shown in **Figure 7.** Different kinds of peppers are displayed in one area. Citrus fruits such as oranges, lemons, and grapefruits are stocked in another area. There are many different ways to organize produce in a grocery store. In a similar way, there have been many different ideas about how to organize, or classify, living things.

A Greek philosopher named Aristotle (384 B.C.–322 B.C.) was one of the first people to classify organisms. Aristotle placed all organisms into two large groups, plants and animals. He classified animals based on the presence of "red blood," the animal's environment, and the shape and size of the animal. He classified plants according to the structure and size of the plant and whether the plant was a tree, a shrub, or an herb.

Figure 7 The produce in this store is classified into groups.

Visual Check What other ways can you think of to classify and organize produce?

Lesson 2
EXPLORE

SCIENCE USE V. COMMON USE

kingdom

Science Use a classification category that ranks above phylum and below domain

Common Use a territory ruled by a king or a queen

Determining Kingdoms

In the 1700s, Carolus Linnaeus, a Swedish physician and botanist, classified organisms based on similar structures. Linnaeus placed all organisms into two main groups, called kingdoms. Over the next 200 years, people learned more about organisms and discovered new organisms. In 1969 American biologist Robert H. Whittaker proposed a five-kingdom system for classifying organisms. His system included kingdoms Monera, Protista, Plantae, Fungi, and Animalia.

Determining Domains

The classification system of living things is still changing. The current classification method is called systematics. Systematics uses all the evidence that is known about organisms to classify them. This evidence includes an organism's cell type, its habitat, the way an organism obtains food and energy, structure and function of its features, and the common ancestry of organisms. Systematics also includes molecular analysis—the study of molecules such as DNA within organisms.

Using systematics, scientists identified two distinct groups in Kingdom Monera—Bacteria and Archaea (ar KEE uh). This led to the development of another level of classification called domains. All organisms are now classified into one of three domains—Bacteria, Archaea, or Eukarya (yew KER ee uh) —and then into one of six kingdoms, as shown in **Table 2**.

 Key Concept Check What evidence is used to classify living things into groups?

Table 2 Domains and Kingdoms						
Domain	Bacteria	Archaea	Eukarya			
Kingdom	Bacteria	Archaea	Protista	Fungi	Plantae	Animalia
Example						
Characteristics	Bacteria are simple unicellular organisms.	Archaea are simple unicellular organisms that often live in extreme environments.	Protists are unicellular and are more complex than bacteria or archaea.	Fungi are unicellular or multicellular and absorb food.	Plants are multicellular and make their own food.	Animals are multicellular and take in their food.

Scientific Names

Suppose you did not have a name. What would people call you? All organisms, just like people, have names. When Linnaeus grouped organisms into kingdoms, he also developed a system for naming organisms. This naming system, called binomial nomenclature (bi NOH mee ul · NOH mun klay chur), is the system we still use today.

Binomial Nomenclature

Linneaus's naming system, **binomial nomenclature,** *gives each organism a two-word scientific name,* such as *Ursus arctos* for a brown bear. This two-word scientific name is the name of an organism's species (SPEE sheez). A **species** *is a group of organisms that have similar traits and are able to produce fertile offspring.* In binomial nomenclature, the first word is the organism's genus (JEE nus) name, such as *Ursus*. A **genus** *is a group of similar species.* The second word might describe the organism's appearance or its behavior.

How do species and genus relate to kingdoms and domains? Similar species are grouped into one genus (plural, genera). Similar genera are grouped into families, then orders, classes, phyla, kingdoms, and finally domains, as shown for the grizzly bear in **Table 3**.

WORD ORIGIN

genus
from Greek *genos*, means "race, kind"

Table 3 The classification of the brown bear or grizzly bear shows that it belongs to the order Carnivora.

Table 3 Classification of the Brown Bear

Taxonomic Group	Number of Species	Examples
Domain Eukarya	About 4–10 million	
Kingdom Animalia	About 2 million	
Phylum Chordata	About 50,000	
Class Mammalia	About 5,000	
Order Carnivora	About 270	
Family Ursidae	8	
Genus *Ursus*	4	
Species *Ursus arctos*	1	

Visual Check What domain does the brown bear belong to?

Uses of Scientific Names

When you talk about organisms, you might use names such as bird, tree, or mushroom. However, these are common names for a number of different species. Sometimes there are several common names for one organism. The animal in **Table 3** on the previous page might be called a brown bear or a grizzly bear, but it has only one scientific name, *Ursus arctos*.

Other times, a common name might refer to several different types of organisms. For example, you might call both of the trees in **Figure 8** pine trees. But these trees are two different species. How can you tell? Scientific names are important for many reasons. Each species has its own scientific name. Scientific names are the same worldwide. This makes communication about organisms more effective because everyone uses the same name for the same species.

Key Concept Check Why does every species have a scientific name?

▲ **Figure 8** These trees are two different species. *Pinus alba* has long needles, and *Tsuga canadensis* has short needles.

Make a horizontal two-tab book to compare two of the tools scientists use to identify organisms—dichotomous keys and cladograms.

Classification Tools

Suppose you go fishing and catch a fish you don't recognize. How could you figure out what type of fish you have caught? There are several tools you can use to identify organisms.

Dichotomous Keys

A **dichotomous key** *is a series of descriptions arranged in pairs that leads the user to the identification of an unknown organism.* The chosen description leads to either another pair of statements or the identification of the organism. Choices continue until the organism is identified. The dichotomous key shown in **Figure 9** identifies several species of fish.

Dichotomous Key

1. a. This fish has a mouth that extends past its eye. It is an arrow goby.	
b. This fish does not have a mouth that extends past its eye. Go to step 2.	
2. a. This fish has a dark body with stripes. It is a chameleon goby.	
b. This fish has a light body with no stripes. Go to step 3.	
3. a. This fish has a black-tipped dorsal fin. It is a bay goby.	
b. This fish has a speckled dorsal fin. It is a yellowfin goby.	

▲ **Figure 9** Dichotomous keys include a series of questions to identify organisms.

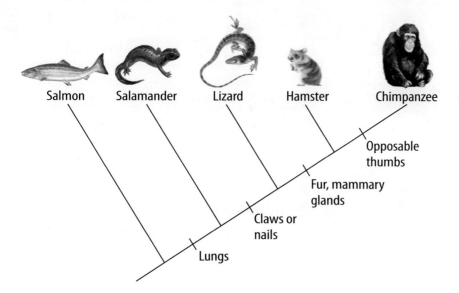

Figure 10 A cladogram shows relationships among species. In this cladogram, salamanders are more closely related to lizards than they are to hamsters.

Concepts in Motion
Animation

Cladograms

A family tree shows the relationships among family members, including common ancestors. Biologists use a similar diagram, called a cladogram. A **cladogram** *is a branched diagram that shows the relationships among organisms, including common ancestors.* A cladogram, as shown in **Figure 10,** has a series of branches. Notice that each branch follows a new characteristic. Each characteristic is observed in all the species to its right. For example, the salamander, lizard, hamster, and chimpanzee have lungs, but the salmon does not. Therefore, they are more closely related to each other than they are to the salmon.

Inquiry MiniLab   20 minutes

How would you name an unknown organism?
Assign scientific names to four unknown alien organisms from a newly discovered planet.

1. Use the table to assign scientific names to identify each alien.
2. Compare your names with those of your classmates.

Analyze and Conclude

1. **Explain** why you chose the two-word names for each organism.
2. **Compare** your names to those of a classmate. Explain any differences.
3. 🔑 **Key Concept** Discuss how two-word scientific names help scientists identify and organize living things.

Prefix	Meaning	Suffix	Meaning
mon–	one	–antennius	antenna
di–	two	–ocularus	eye
rectanguli–	square	–formus	shape
trianguli–	triangle	–uris	tail

Lesson 2
EXPLAIN

Lesson 2 Review

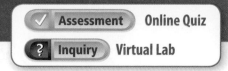

Visual Summary

All organisms are classified into one of three domains: Bacteria, Archaea, or Eukarya.

Every organism has a unique species name.

A dichotomous key helps to identify an unknown organism through a series of paired descriptions.

FOLDABLES

Use your lesson Foldable to review the lesson. Save your Foldable for the project at the end of the chapter.

What do you think NOW?

You first read the statements below at the beginning of the chapter.

3. A dichotomous key can be used to identify an unknown organism.

4. Physical similarities are the only traits used to classify organisms.

Did you change your mind about whether you agree or disagree with the statements? Rewrite any false statements to make them true.

Use Vocabulary

1. A naming system that gives every organism a two-word name is _____ _____.

2. **Use the term** *dichotomous key* in a sentence.

3. **Organisms** of the same _____ are able to produce fertile offspring.

Understand Key Concepts

4. **Describe** how you write a scientific name.

5. **Compare** the data available today on how to classify things with the data available during Aristotle's time.

6. Which is NOT used to classify organisms?
    A. ancestry
    B. habitat
    C. age of the organism
    D. molecular evidence

Interpret Graphics

7. **Organize Information** Copy and fill in the graphic organizer below to show how organisms are classified.

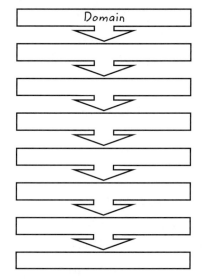

Critical Thinking

8. **Suggest** a reason scientists might consider changing the current classification system.

9. **Evaluate** the importance of scientific names.

Inquiry Skill Practice: Classify

20 minutes

How can you identify a beetle?

A dichotomous key is one of the tools scientists use to identify an unknown organism and **classify** it into a group. To use a dichotomous key, a scientist examines specific characteristics of the unknown organism and compares them to characteristics of known organisms.

Learn It

Sorting objects or events into groups based on common features is called classifying. When classifying, select one feature that is shared by some members of the group, but not by all. Place those members that share the feature in a subgroup. You can **classify** objects or events into smaller and smaller subgroups based on characteristics.

Try It

1. Use the dichotomous key to identify beetle A. Choose between the first pair of descriptions. Follow the instructions for the next choice. Notice that each description either ends in the name of the beetle or instructs you to go on to another set of choices.

2. In your Science Journal, record the identity of the beetle using both its common name and scientific name.

3. Repeat steps 1 and 2 for beetles B, C, and D.

Apply It

4. Think about the choices in each step of the dichotomous key. What conclusion can be made if you arrive at a step and neither choice seems correct?

5. Predict whether a dichotomous key will work if you start at a location other than the first description. Support your reasoning.

6. **Key Concept** How did the dichotomous key help you classify the unknown beetles?

Dichotomous Key

1A.	The beetle has long, thin antennae. Go to 5.
1B.	The beetle does not have long, thin antennae. Go to 2.
2A.	The beetle has short antennae that branch. Go to 3.
2B.	The beetle does not have short antennae that branch. It is a stag beetle, *Lucanus cervus*.
3A.	The beetle has a triangular structure between wing covers and upper body. It is a Japanese beetle, *Popillia japonica*.
3B.	The beetle does not have a triangular structure. Go to 4.
4A.	The beetle has a wide, rounded body. It is a June bug, *Cotinis nitida*.
4B.	The beetle does not have a wide, rounded body. It is a death watch beetle, *Xestobium rufovillosum*.
5A.	The beetle has a distinct separation between body parts. Go to 6.
5B.	The beetle has no distinct separation between body parts. It is a firefly, *Photinus pyralis*.
6A.	The beetle has a black, gray, and white body with two black eyespots. It is an eyed click beetle, *Alaus oculatis*.
6B.	The beetle has a dull brown body with light stripes. It is a click beetle, *Chalcolepidius limbatus*.

Lesson 2
EXTEND

Lesson 3
Exploring Life

Reading Guide

Key Concepts 🔑
ESSENTIAL QUESTIONS

- How did microscopes change our ideas about living things?
- What are the types of microscopes, and how do they compare?

Vocabulary
light microscope p. 28
compound microscope p. 28
electron microscope p. 29

 Multilingual eGlossary

Inquiry Giant Insect?

Although this might look like a giant insect, it is a photo of a small tick taken with a high-powered microscope. This type of microscope can enlarge an image of an object up to 200,000 times. How can seeing an enlarged image of a living thing help you understand life?

Launch Lab

15 minutes

Can a water drop make objects appear bigger or smaller?

For centuries, people have been looking for ways to see objects in greater detail. How can something as simple as a drop of water make this possible?

1. Read and complete a lab safety form.
2. Lay a sheet of **newspaper** on your desk. Examine a line of text, noting the size and shape of each letter. Record your observations in your Science Journal.
3. Add a large drop of **water** to the center of a piece of **clear plastic.** Hold the plastic about 2 cm above the same line of text.
4. Look through the water at the line of text you viewed in step 2. Record your observations.

Think About This

1. Describe how the newsprint appeared through the drop of water.
2. **Key Concept** How might microscopes change your ideas about living things?

The Development of Microscopes

Have you ever used a magnifying lens to see details of an object? If so, then you have used a tool similar to the first microscope. The invention of microscopes enabled people to see details of living things that they could not see with the unaided eye. The microscope also enabled people to make many discoveries about living things.

In the late 1600s the Dutch merchant Anton van Leeuwenhoek (LAY vun hook) made one of the first microscopes. His microscope, similar to the one shown in **Figure 11,** had one lens and could magnify an image about 270 times its original size. Another inventor of microscopes was Robert Hooke. In the early 1700s Hooke made one of the most significant discoveries using a microscope. He observed and named cells. Before microscopes, people did not know that living things are made of cells.

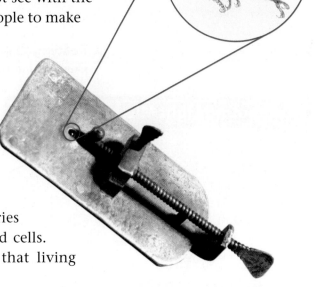

Figure 11 Anton van Leeuwenhoek observed pond water and insects using a microscope like the one shown above.

Key Concept Check How did microscopes change our ideas about living things?

Lesson 3

EXPLORE

Math Skills

Use Multiplication
The magnifying power of a lens is expressed by a number and a multiplication symbol (×). For example, a lens that makes an object look ten times larger has a power of 10×. To determine a microscope's magnification, multiply the power of the ocular lens by the power of the objective lens. A microscope with a 10× ocular lens and a 10× objective lens magnifies an object 10 × 10, or 100 times.

Practice
What is the magnification of a compound microscope with a 10× ocular lens and a 4× objective lens?

- Math Practice
- Personal Tutor

Types of Microscopes

One characteristic of all microscopes is that they magnify objects. Magnification makes an object appear larger than it really is. Another characteristic of microscopes is resolution—how clearly the magnified object can be seen. The two main types of microscopes—light microscopes and electron microscopes—differ in magnification and resolution.

Light Microscopes

If you have used a microscope in school, then you have probably used a light microscope. **Light microscopes** *use light and lenses to enlarge an image of an object.* A simple light microscope has only one lens. *A light microscope that uses more than one lens to magnify an object is called a* **compound microscope.** A compound microscope magnifies an image first by one lens, called the objective lens. The image is then further magnified by another lens, called the ocular lens. The total magnification of the image is equal to the magnifications of the ocular lens and the objective lens multiplied together.

Light microscopes can enlarge images up to 1,500 times their original size. The resolution of a light microscope is about 0.2 micrometers (μm), or two-millionths of a meter. A resolution of 0.2 μm means you can clearly see points on an object that are at least 0.2 μm apart.

Light microscopes can be used to view living or nonliving objects. In some light microscopes, an object is placed directly under the microscope. For other light microscopes, an object must be mounted on a slide. In some cases, the object, such as the white blood cells in **Figure 12,** must be stained with a dye in order to see any details.

 Reading Check What are some ways an object can be examined under a light microscope?

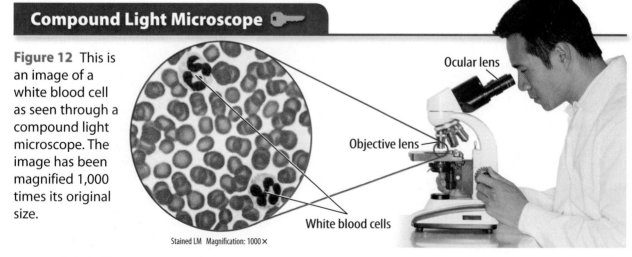

Figure 12 This is an image of a white blood cell as seen through a compound light microscope. The image has been magnified 1,000 times its original size.

Stained LM Magnification: 1000×

Electron Microscopes

You might know that electrons are tiny particles inside atoms. **Electron microscopes** *use a magnetic field to focus a beam of electrons through an object or onto an object's surface.* An electron microscope can magnify an image up to 100,000 times or more. The resolution of an electron microscope can be as small as 0.2 nanometers (nm), or two-billionths of a meter. This resolution is up to 1,000 times greater than a light microscope. The two main types of electron microscopes are transmission electron microscopes (TEMs) and scanning electron microscopes (SEMs).

TEMs are usually used to study extremely small things such as cell structures. Because objects must be mounted in plastic and then very thinly sliced, only dead organisms can be viewed with a TEM. In a TEM, electrons pass through the object and a computer produces an image of the object. A TEM image of a white blood cell is shown in **Figure 13**.

SEMs are usually used to study an object's surface. In an SEM, electrons bounce off the object and a computer produces a three-dimensional image of the object. An image of a white blood cell from an SEM is shown in **Figure 13**. Note the difference in detail in this image compared to the image in **Figure 12** of a white blood cell from a light microscope.

 Key Concept Check What are the types of microscopes, and how do they compare?

REVIEW VOCABULARY
atom
the building block of matter that is composed of protons, neutrons, and electrons

Make a two-column folded chart. Label the front *Types of Microscopes,* and label the inside as shown. Use it to organize your notes about microscopes.

Figure 13 A TEM greatly magnifies thin slices of an object. An SEM is used to view a three-dimensional image of an object.

Electron Microscopes

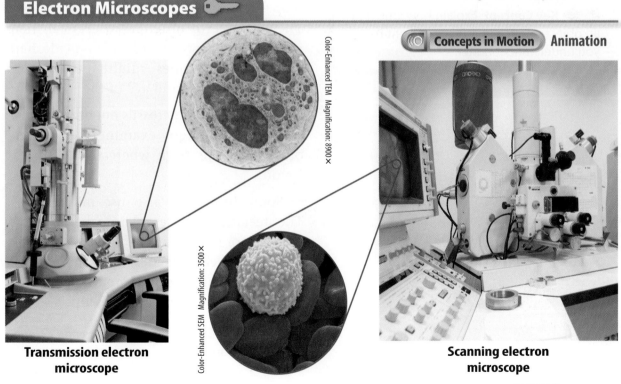

Transmission electron microscope

Scanning electron microscope

Inquiry MiniLab

20 minutes

How do microscopes help us compare living things?

A microscope enables scientists to study objects in greater detail than is possible with the unaided eye. Compare what objects look like with the unaided eye to those same objects observed using a microscope.

1. Read and complete a lab safety form.
2. Examine a **sea sponge,** a **leaf,** and **salt crystals.** Draw each object in your Science Journal.
3. Observe **microscope slides of each object** using a **microscope** on low power.
4. Draw each object as it appears under low power.

Analyze and Conclude

1. **Compare** your sketches of the objects observed with your unaided eye and observed with a microscope.
2. **Key Concept** Explain how studying an object under a microscope might help you understand it better.

Word Origin

microscope
from Latin *microscopium,* means "an instrument for viewing what is small"

Academic Vocabulary

identify
(verb) to determine the characteristics of a person or a thing

Using Microscopes

The microscopes used today are more advanced than the microscopes used by Leeuwenhoek and Hooke. The quality of today's light microscopes and the invention of electron microscopes have made the microscope a useful tool in many fields.

Health Care

People in health-care fields, such as doctors and laboratory technicians, often use microscopes. Microscopes are used in surgeries, such as cataract surgery and brain surgery. They enable doctors to view the surgical area in greater detail. The area being viewed under the microscope can also be displayed on a TV monitor so that other people can watch the procedure. Laboratory technicians use microscopes to analyze body fluids, such as blood and urine. They also use microscopes to determine whether tissue samples are healthy or diseased.

Other Uses

Health care is not the only field that uses microscopes. Have you ever wondered how police determine how and where a crime happened? Forensic scientists use microscopes to study evidence from crime scenes. The presence of different insects can help identify when and where a homicide happened. Microscopes might be used to identify the type and age of the insects.

People who study fossils might use microscopes. They might examine a fossil and other materials from where the fossil was found.

Some industries also use microscopes. The steel industry uses microscopes to examine steel for impurities. Microscopes are used to study jewels and identify stones. Stones have some markings and impurities that can be seen only by using a microscope.

 Reading Check List some uses of microscopes.

Lesson 3 Review

Assessment Online Quiz

Visual Summary

Living organisms can be viewed with light microscopes.

A compound microscope is a type of light microscope that has more than one lens.

Living organisms cannot be viewed with a transmission electron microscope.

Use your lesson Foldable to review the lesson. Save your Foldable for the project at the end of the chapter.

What do you think NOW?

You first read the statements below at the beginning of the chapter.

5. Most cells are too small to be seen with the unaided eye.

6. Only scientists use microscopes.

Did you change your mind about whether you agree or disagree with the statements? Rewrite any false statements to make them true.

Use Vocabulary

1. **Define** the term *light microscope* in your own words.

2. A(n) _____ focuses a beam of electrons through an object or onto an object's surface.

Understand Key Concepts

3. **Explain** how the discovery of microscopes has changed what we know about living things.

4. Which microscope would you use if you wanted to study the surface of an object?
    A. compound microscope
    B. light microscope
    C. scanning electron microscope
    D. transmission electron microscope

Interpret Graphics

5. **Identify** Copy and fill in the graphic organizer below to identify four uses of microscopes.

6. **Compare** the images of the white blood cells below. How do they differ?

Critical Thinking

7. **Develop** a list of guidelines for choosing a microscope to use.

Math Skills

— Math Practice —

8. A student observes a blood sample with a compound microscope that has a 10× ocular lens and a 40× objective lens. How much larger do the blood cells appear under the microscope?

Lesson 3 • **31**
EVALUATE

Inquiry Lab

45 minutes

Materials

a collection of objects

Constructing a Dichotomous Key

A dichotomous key is a series of descriptions arranged in pairs. Each description leads you to the name of the object or to another set of choices until you have identified the organism. In this lab, you will create a dichotomous key to classify objects.

Question

How can you create a dichotomous key to identify objects?

Procedure

1. Read and complete a lab safety form.
2. Obtain a container of objects from your teacher.
3. Examine the objects, and then brainstorm a list of possible characteristics. You might look at each object's size, shape, color, odor, texture, or function.
4. Choose a characteristic that would separate the objects into two groups. Separate the objects based on whether or not they have this characteristic. This characteristic will be used to begin a dichotomous key, like the example below.

Dichotomous Key to Identify Office Supplies

| The object is made of wood. Go to 1. |
| The object is not made of wood. Go to 2. |
| 1. The object is longer than 20 cm. Go to 5. |
| 3. The object is not longer than 20 cm. Go to 9. |
| 2. The object is made of metal. Go to 6. |
| 4. The object is not made of metal. Go to 10. |

32

5. Write a sentence to describe the characteristic in step 4, and then write "Go to 1." Write another sentence that has the word "not" in front of the characteristic. Then write "Go to 2."

6. Repeat steps 4 and 5 for the two new groups. Give sentences for new groups formed from the first group consecutive odd numbers. Give sentences for groups formed from the second group consecutive even numbers. Remember to add the appropriate "Go to" directions.

7. Repeat steps 4–6 until there is one object in each group. Give each object an appropriate two-word name.

8. Give your collection of objects and your dichotomous key to another group. Have them identify each object using your dichotomous key. Have them record their answers.

Analyze and Conclude

9. **Evaluate** Was the other team able to correctly identify the collection of objects using your dichotomous key? Why or why not?

10. **The Big Idea** Summarize how dichotomous keys are useful in identifying unknown objects.

Lab Tips

☑ Base the questions in your key on observable, measurable, or countable characteristics. Avoid questions that refer to how something is used or how you think or feel about an item.

☑ Remember to start with general questions and then get more and more specific.

Communicate Your Results

Create a poster using drawings or photos of each object you identified. Include your two-word names for the objects.

Teach a peer how to use a dichotomous key. Let the peer use your collection to have a first-hand experience with how a key works.

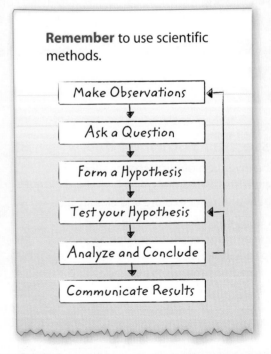

Remember to use scientific methods.

- Make Observations
- Ask a Question
- Form a Hypothesis
- Test your Hypothesis
- Analyze and Conclude
- Communicate Results

Chapter 1 Study Guide

All living things have certain characteristics in common and can be classified using several methods. The invention of the microscope has enabled us to explore life further, which has led to changes in classification.

Key Concepts Summary

Lesson 1: Characteristics of Life

- An **organism** is classified as a living thing because it has all the characteristics of life.
- All living things are organized, grow and develop, reproduce, respond to stimuli, maintain **homeostasis,** and use energy.

Vocabulary

organism p. 9
cell p. 10
unicellular p. 10
multicellular p. 10
homeostasis p. 13

Lesson 2: Classifying Organisms

- Living things are classified into different groups based on physical or molecular similarities.
- Some **species** are known by many different common names. To avoid confusion, every species has a scientific name based on a system called **binomial nomenclature.**

binomial nomenclature p. 21
species p. 21
genus p. 21
dichotomous key p. 22
cladogram p. 23

Lesson 3: Exploring Life

- The invention of microscopes allowed scientists to view cells, which enabled them to further explore and classify life.
- A **light microscope** uses light and has one or more lenses to enlarge an image up to about 1,500 times its original size. An **electron microscope** uses a magnetic field to direct beams of electrons, and it enlarges an image 100,000 times or more.

light microscope p. 28
compound microscope p. 28
electron microscope p. 29

Study Guide

- Personal Tutor
- Vocabulary eGames
- Vocabulary eFlashcards

FOLDABLES Chapter Project

Assemble your lesson Foldables as shown to make a Chapter Project. Use the project to review what you have learned in this chapter.

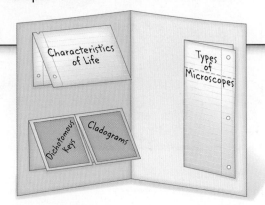

Use Vocabulary

1. A(n) _____ organism is made of only one cell.
2. Something with all the characteristics of life is a(n) _____.
3. A(n) _____ shows the relationships among species.
4. A group of similar species is a(n) _____.
5. A(n) _____ has a resolution up to 1,000 times greater than a light microscope.
6. A(n) _____ is a light microscope that uses more than one lens to magnify an image.

 Interactive Concept Map

Link Vocabulary and Key Concepts

Copy this concept map, and then use vocabulary terms from the previous page to complete the concept map.

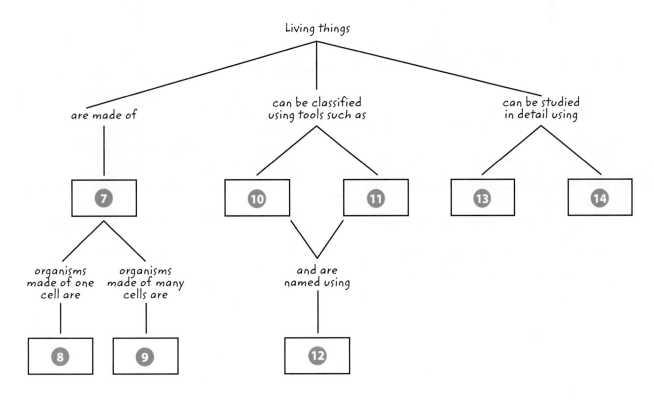

Chapter 1 Study Guide • 35

Chapter 1 Review

Understand Key Concepts

1. Which is an internal stimulus?
   A. an increase in moisture
   B. feelings of hunger
   C. number of hours of daylight
   D. the temperature at night

2. Which is an example of growth and development?
   A. a caterpillar becoming a butterfly
   B. a chicken laying eggs
   C. a dog panting
   D. a rabbit eating carrots

3. Based on the food web below, what is an energy source for the mouse?

   A. fox
   B. grass
   C. owl
   D. snake

4. Which shows the correct order for the classification of species?
   A. domain, kingdom, class, order, phylum, family, genus, species
   B. domain, kingdom, phylum, class, order, family, genus, species
   C. domain, kingdom, phylum, class, order, family, species, genus
   D. domain, kingdom, phylum, order, class, family, genus, species

5. The organism shown below belongs in which kingdom?

   A. Animalia
   B. Archaea
   C. Bacteria
   D. Plantae

6. Which was discovered using a microscope?
   A. blood
   B. bones
   C. cells
   D. hair

7. What type of microscope would most likely be used to obtain an image of a live roundworm?
   A. compound light microscope
   B. scanning electron microscope
   C. simple light microscope
   D. transmission electron microscope

8. Which best describes a compound microscope?
   A. uses electrons to magnify the image of an object
   B. uses multiple lenses to magnify the image of an object
   C. uses one lens to magnify the image of an object
   D. uses sound waves to magnify the image of an object

Chapter Review

Assessment
Online Test Practice

Critical Thinking

9 Distinguish between a unicellular organism and a multicellular organism.

10 Critique the following statement: An organism that is made of only one cell does not need organization.

11 Infer In the figure below, which plant is responding to a lack of water in its environment? Explain your answer.

12 Explain how using a dichotomous key can help you identify an organism.

13 Describe how the branches on a cladogram show the relationships among organisms.

14 Assess the effect of molecular evidence on the classification of organisms.

15 Compare light microscopes and electron microscopes.

16 State how microscopes have changed the way living things are classified.

17 Compare magnification and resolution.

18 Evaluate the impact microscopes have on our daily lives.

Writing in Science

19 Write a five-sentence paragraph explaining the importance of scientific names. Be sure to include a topic sentence and a concluding sentence in your paragraph.

REVIEW THE BIG IDEA

20 Define the characteristics that all living things share.

21 The photo below shows living and nonliving things. How would you classify the living things by domain and kingdom?

Math Skills

Review Math Practice

Use Multiplication

22 A microscope has an ocular lens with a power of 5× and an objective lens with a power of 50×. What is the total magnification of the microscope?

23 A student observes a unicellular organism with a microscope that has a 10× ocular lens and a 100× objective lens. How much larger does the organism look through this microscope?

24 The ocular lens on a microscope has a power of 10×. The microscope makes objects appear 500 times larger. What is the power of the objective lens?

Chapter 1 Review • 37

Standardized Test Practice

Record your answers on the answer sheet provided by your teacher or on a sheet of paper.

Multiple Choice

1. What feature of living things do the terms *unicellular* and *multicellular* describe?
   A how they are organized
   B how they reproduce
   C how they maintain temperature
   D how they produce macromolecules

Use the diagram below to answer question 2.

2. Which characteristic of life does the diagram show?
   A homeostasis
   B organization
   C growth and development
   D response to stimuli

3. A newly discovered organism is 1 m tall, multicellular, green, and it grows on land and performs photosynthesis. To which kingdom does it most likely belong?
   A Animalia
   B Fungi
   C Plantae
   D Protista

4. Unicellular organisms are members of which kingdoms?
   A Animalia, Archaea, Plantae
   B Archaea, Bacteria, Protista
   C Bacteria, Fungi, Plantae
   D Fungi, Plantae, Protista

5. Which microscope would best magnify the outer surface of a cell?
   A compound light
   B scanning electron
   C simple dissecting
   D transmission electron

Use the diagram below to answer question 6.

6. Which discovery was NOT made with the instrument above?
   A Bacterial cells have thick walls.
   B Blood is a mixture of components.
   C Insects have small body parts.
   D Tiny organisms live in pond water.

7. Which statement is false?
   A Binomial names are given to all known organisms.
   B Binomial names are less precise than common names.
   C Binomial names differ from common names.
   D Binomial names enable scientists to communicate accurately.

Standardized Test Practice

Use the diagram below to answer question 8.

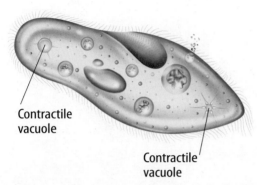

Contractile vacuole

Contractile vacuole

8 Which is the function of the structures in this paramecium?
- A growth
- B homeostasis
- C locomotion
- D reproduction

9 Which sequence is from the smallest group of organisms to the largest group of organisms?
- A genus → family → species
- B genus → species → family
- C species → family → genus
- D species → genus → family

10 Which information about organisms is excluded in the study of systematics?
- A calendar age
- B molecular analysis
- C energy source
- D normal habitat

Constructed Response

11 Copy and complete the table below about the six characteristics of life.

Characteristic	Explanation

12 Choose one characteristic of living things and explain how it affects everyday human life. From your own knowledge, give a specific example.

Use the diagram below to answer question 13.

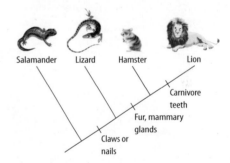

Salamander Lizard Hamster Lion
Carnivore teeth
Fur, mammary glands
Claws or nails

13 Explain why the lion is more closely related to the hamster than the hamster is related to the salamander.

NEED EXTRA HELP?													
If You Missed Question...	1	2	3	4	5	6	7	8	9	10	11	12	13
Go to Lesson...	1	1	2	2	3	3	2	1	2	2	1	1	2

Chapter 2
Cell Structure and Function

THE BIG IDEA How do the structures and processes of a cell enable it to survive?

Inquiry Alien Life?

You might think this unicellular organism looks like something out of a science-fiction movie. Although it looks scary, the hairlike structures in its mouth enable the organism to survive.

- What do you think the hairlike structures do?
- How might the shape of the hairlike structures relate to their function?
- How do you think the structures and processes of a cell enable it to survive?

Get Ready to Read

What do you think?
Before you read, decide if you agree or disagree with each of these statements. As you read this chapter, see if you change your mind about any of the statements.

1. Nonliving things have cells.
2. Cells are made mostly of water.
3. Different organisms have cells with different structures.
4. All cells store genetic information in their nuclei.
5. Diffusion and osmosis are the same process.
6. Cells with large surface areas can transport more than cells with smaller surface areas.
7. ATP is the only form of energy found in cells.
8. Cellular respiration occurs only in lung cells.

ConnectED Your one-stop online resource

connectED.mcgraw-hill.com

- Video
- Audio
- Review
- Inquiry
- WebQuest
- Assessment
- Concepts in Motion
- Multilingual eGlossary

Lesson 1

Cells and Life

Reading Guide

Key Concepts 🔑
ESSENTIAL QUESTIONS

- How did scientists' understanding of cells develop?
- What basic substances make up a cell?

Vocabulary
cell theory p. 44
macromolecule p. 45
nucleic acid p. 46
protein p. 47
lipid p. 47
carbohydrate p. 47

 Multilingual eGlossary

Inquiry Two of a Kind?

At first glance, the plant and animal in the photo might seem like they have nothing in common. The plant is rooted in the ground, and the iguana can move quickly. Are they more alike than they appear? How can you find out?

Inquiry Launch Lab

10 minutes

What's in a cell?

Most plants grow from seeds. A seed began as one cell, but a mature plant can be made up of millions of cells. How does a seed change and grow into a mature plant?

1. Read and complete a lab safety form.
2. Use a **toothpick** to gently remove the thin outer covering of a **bean seed** that has soaked overnight.
3. Open the seed with a **plastic knife,** and observe its inside with a **magnifying lens.** Draw the inside of the seed in your Science Journal.
4. Gently remove the small, plantlike embryo, and weigh it on a **balance.** Record its mass in your Science Journal.
5. Gently pull a **bean seedling** from the soil. Rinse the soil from the roots. Weigh the seedling, and record the mass.

Think About This

1. How did the mass of the embryo and the bean seedling differ?
2. **Key Concept** If a plant begins as one cell, where do all the cells come from?

Understanding Cells

Have you ever looked up at the night sky and tried to find other planets in our solar system? It is hard to see them without using a telescope. This is because the other planets are millions of kilometers away. Just like we can use telescopes to see other planets, we can use microscopes to see the basic units of all living things—cells. But people didn't always know about cells. Because cells are so small, early scientists had no tools to study them. It took hundreds of years for scientists to learn about cells.

More than 300 years ago, an English scientist named Robert Hooke built a microscope. He used the microscope to look at cork, which is part of a cork oak tree's bark. What he saw looked like the openings in a honeycomb, as shown in **Figure 1.** The openings reminded him of the small rooms, called cells, where monks lived. He called the structures cells, from the Latin word *cellula* (SEL yuh luh), which means "small rooms."

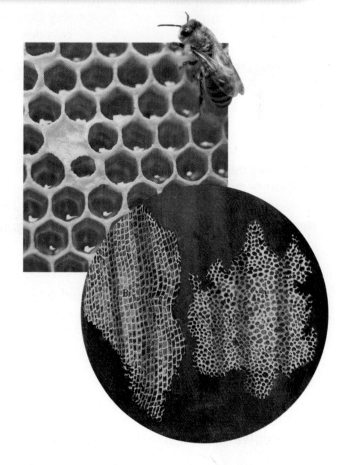

Figure 1 To Robert Hooke, the cells of cork looked like the openings in a honeycomb.

Lesson 1
EXPLORE

The Cell Theory

After Hooke's discovery, other scientists began making better microscopes and looking for cells in many other places, such as pond water and blood. The newer microscopes enabled scientists to see different structures inside cells. Matthias Schleiden (SHLI dun), a German scientist, used one of the new microscopes to look at plant cells. Around the same time, another German scientist, Theodor Schwann, used a microscope to study animal cells. Schleiden and Schwann realized that plant and animal cells have similar features. You'll read about many of these features in Lesson 2.

Almost two decades later, Rudolf Virchow (VUR koh), a German doctor, proposed that all cells come from preexisting cells, or cells that already exist. The observations made by Schleiden, Schwann, and Virchow were combined into one theory. As illustrated in **Table 1**, *the* **cell theory** *states that all living things are made of one or more cells, the cell is the smallest unit of life, and all new cells come from preexisting cells.* After the development of the cell theory, scientists raised more questions about cells. If all living things are made of cells, what are cells made of?

Key Concept Check How did scientists' understanding of cells develop?

> **REVIEW VOCABULARY**
> **theory**
> explanation of things or events based on scientific knowledge resulting from many observations and experiments

Table 1 Scientists developed the cell theory after studying cells with microscopes.

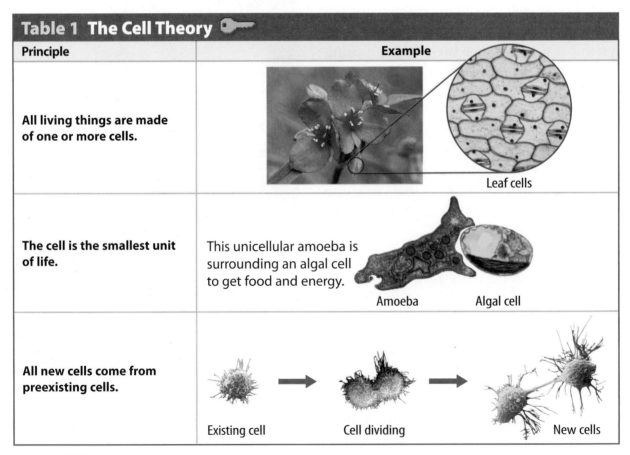

Table 1 The Cell Theory	
Principle	**Example**
All living things are made of one or more cells.	Leaf cells
The cell is the smallest unit of life.	This unicellular amoeba is surrounding an algal cell to get food and energy. Amoeba — Algal cell
All new cells come from preexisting cells.	Existing cell → Cell dividing → New cells

Basic Cell Substances

Have you ever watched a train travel down a railroad track? The locomotive pulls train cars that are hooked together. Like a train, many of the substances in cells are made of smaller parts that are joined together. *These substances, called* **macromolecules,** *form by joining many small molecules together.* As you will read later in this lesson, macromolecules have many important roles in cells. But macromolecules cannot function without one of the most important substances in cells—water.

The Main Ingredient—Water

The main ingredient in any cell is water. It makes up more than 70 percent of a cell's volume and is essential for life. Why is water such an important molecule? In addition to making up a large part of the inside of cells, water also surrounds cells. The water surrounding your cells helps to insulate your body, which maintains homeostasis, or a stable internal environment.

The structure of a water molecule makes it ideal for dissolving many other substances. Substances must be in a liquid to move into and out of cells. A water molecule has two areas:

- An area that is more negative (−), called the negative end; this end can attract the positive part of another substance.

- An area that is more positive (+), called the positive end; this end can attract the negative part of another substance.

Examine **Figure 2** to see how the positive and negative ends of water molecules dissolve salt crystals.

WORD ORIGIN

macromolecule
from Greek *makro–*, means "long"; and Latin *molecula*, means "mass"

Figure 2 The positive and negative ends of a water molecule attract the positive and negative parts of another substance, similar to the way magnets are attracted to each other.

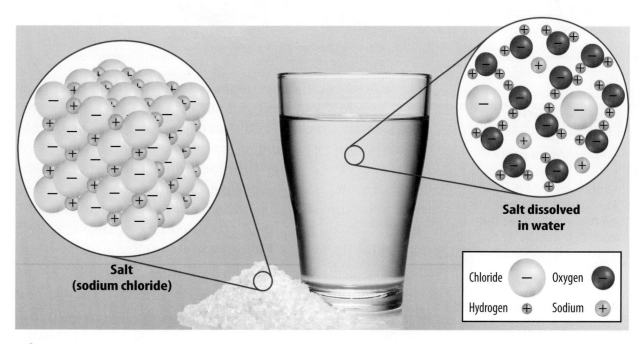

Salt (sodium chloride)

Salt dissolved in water

Chloride — Oxygen —
Hydrogen + Sodium +

Visual Check Which part of the salt crystal is attracted to the oxygen in the water molecule?

Macromolecules

Although water is essential for life, all cells contain other substances that enable them to function. Recall that macromolecules are large molecules that form when smaller molecules join together. As shown in **Figure 3,** there are four types of macromolecules in cells: nucleic acids, proteins, lipids, and carbohydrates. Each type of macromolecule has unique functions in a cell. These functions range from growth and communication to movement and storage.

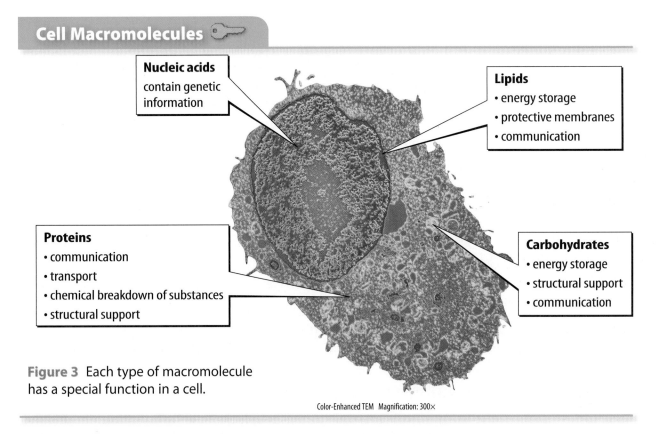

Figure 3 Each type of macromolecule has a special function in a cell.

Color-Enhanced TEM Magnification: 300×

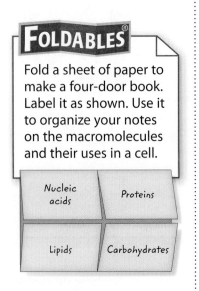

Fold a sheet of paper to make a four-door book. Label it as shown. Use it to organize your notes on the macromolecules and their uses in a cell.

Nucleic Acids Both deoxyribonucleic (dee AHK sih ri boh noo klee ihk) acid (DNA) and ribonucleic (ri boh noo KLEE ihk) acid (RNA) are nucleic acids. **Nucleic acids** *are macromolecules that form when long chains of molecules called nucleotides* (NEW klee uh tidz) *join together.* The order of nucleotides in DNA and RNA is important. If you change the order of words in a sentence, you can change the meaning of the sentence. In a similar way, changing the order of nucleotides in DNA and RNA can change the genetic information in a cell.

Nucleic acids are important in cells because they contain genetic information. This information can pass from parents to offspring. DNA includes instructions for cell growth, cell reproduction, and cell processes that enable a cell to respond to its environment. DNA is used to make RNA. RNA is used to make proteins.

Proteins The macromolecules necessary for nearly everything cells do are proteins. **Proteins** *are long chains of amino acid molecules.* You just read that RNA is used to make proteins. RNA contains instructions for joining amino acids together.

Cells contain hundreds of proteins. Each protein has a unique function. Some proteins help cells communicate with each other. Other proteins transport substances around inside cells. Some proteins, such as amylase (AM uh lays) in saliva, help break down nutrients in food. Other proteins, such as keratin (KER uh tun)—a protein found in hair, horns, and feathers—provide structural support.

Lipids Another group of macromolecules found in cells is lipids. *A* **lipid** *is a large macromolecule that does not dissolve in water.* Because lipids do not mix with water, they play an important role as protective barriers in cells. They are also the major part of cell membranes. Lipids play roles in energy storage and in cell communication. Examples of lipids are cholesterol (kuh LES tuh rawl), phospholipids (fahs foh LIH pids), and vitamin A.

 Reading Check Why are lipids important to cells?

Carbohydrates *One sugar molecule, two sugar molecules, or a long chain of sugar molecules make up* **carbohydrates** (kar boh HI drayts). Carbohydrates store energy, provide structural support, and are needed for communication between cells. Sugars and starches are carbohydrates that store energy. Fruits contain sugars. Breads and pastas are mostly starch. The energy in sugars and starches can be released quickly through chemical reactions in cells. Cellulose is a carbohydrate in the cell walls in plants that provides structural support.

 Key Concept Check What basic substances make up a cell?

Inquiry MiniLab — 25 minutes

How can you observe DNA?

Nucleic acids are macromolecules that are important in cells because they contain an organism's genetic information. In this lab, you will observe one type of nucleic acid, DNA, in onion root-tip cells using a compound light microscope.

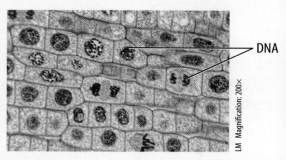

1. Read and complete a lab safety form.
2. Obtain a **microscope** and a **slide** from your teacher. Use care and properly handle your microscope.
3. Observe the **onion root-tip cells** at the magnifications assigned by your teacher.
4. Determine the approximate number of cells in your field of view and the number of cells with visible DNA. Record these numbers in your Science Journal.

Analyze and Conclude

1. **Calculate** Using your data, find the percentage of cells with visible DNA that you saw in your microscope's field of view.
2. **Compare** your results with the results of other students. Are all the results the same? Explain.
3. **Create** a data table for the entire class that lists individual results.
4. **Calculate** the total percentage of cells with visible DNA at each magnification.
5. **Key Concept** Did looking at the cells at different magnifications change the percentage of cells with visible DNA? Explain.

Lesson 1 Review

Visual Summary

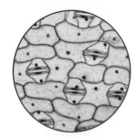
The cell theory summarizes the main principles for understanding that the cell is the basic unit of life.

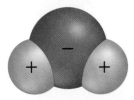
Water is the main ingredient in every cell.

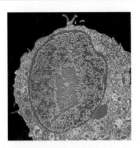
A nucleic acid, such as DNA, contains the genetic information for a cell.

FOLDABLES
Use your lesson Foldable to review the lesson. Save your Foldable for the project at the end of the chapter.

What do you think NOW?

You first read the statements below at the beginning of the chapter.

1. Nonliving things have cells.
2. Cells are made mostly of water.

Did you change your mind about whether you agree or disagree with the statements? Rewrite any false statements to make them true.

Use Vocabulary

1. The _____ _____ states that the cell is the basic unit of all living things.
2. **Distinguish** between a carbohydrate and a lipid.
3. **Use the term** *nucleic acid* in a sentence.

Understand Key Concepts

4. Which macromolecule is made from amino acids?
   A. lipid
   B. protein
   C. carbohydrate
   D. nucleic acid
5. **Describe** how the invention of the microscope helped scientists understand cells.
6. **Compare** the functions of DNA and proteins in a cell.

Interpret Graphics

7. **Summarize** Copy and fill in the graphic organizer below to summarize the main principles of the cell theory.

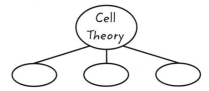

8. **Analyze** How does the structure of the water molecule shown below enable it to interact with other water molecules?

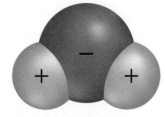

Critical Thinking

9. **Summarize** the functions of lipids in cells.
10. **Hypothesize** why carbohydrates are found in plant cell walls.

A Very Powerful Microscope

How NATURE WORKS

Using technology to look inside cells

If Robert Hooke had used an atomic force microscope (AFM), he would have observed more than just cells. He would have seen the macromolecules inside them! An AFM can scan objects that are only nanometers in size. A nanometer is one one-billionth of a meter. That's 100,000 times smaller than the width of a human hair. AFM technology has enabled scientists to better understand how cells function. It also has given them a three-dimensional look at the macromolecules that make life possible. This is how it works.

Photodiode

2 The cantilever can bend up and down, similar to the way a diving board can bend, in response to pushing and pulling forces between the atoms in the tip and the atoms in the sample.

3 A laser beam senses the cantilever's up and down movements. A computer converts these movements into an image of the sample's surface.

1 A probe moves across a sample's surface to identify the sample's features. The probe consists of a cantilever with a tiny, sharp tip. The tip is about 20 nm in diameter at its base.

It's Your Turn

RESEARCH NASA's Phoenix Mars Lander included an atomic force microscope. Find out what scientists discovered on Mars with this instrument.

Lesson 1 EXTEND

Lesson 2

The Cell

Reading Guide

Key Concepts 🔑
ESSENTIAL QUESTIONS

- How are prokaryotic cells and eukaryotic cells similar, and how are they different?
- What do the structures in a cell do?

Vocabulary

cell membrane p. 52
cell wall p. 52
cytoplasm p. 53
cytoskeleton p. 53
organelle p. 54
nucleus p. 55
chloroplast p. 57

g Multilingual eGlossary

Video BrainPOP®

Inquiry Hooked Together?

What do you think happens when one of the hooks in the photo above goes through one of the loops? The two sides fasten together. The shapes of the hooks and loops in the hook-and-loop tape are suited to their function—to hold the two pieces together.

Inquiry Launch Lab

10 minutes

Why do eggs have shells?

Bird eggs have different structures, such as a shell, a membrane, and a yolk. Each structure has a different function that helps keep the egg safe and assists in development of the baby bird inside of it.

1. Read and complete a lab safety form.
2. Place an **uncooked egg** in a bowl.
3. Feel the shell, and record your observations in your Science Journal.
4. Crack open the egg. Pour the contents into the bowl.
5. Observe the inside of the shell and the contents of the bowl. Record your observations in your Science Journal.

Think About This

1. What do you think is the role of the eggshell?
2. Are there any structures in the bowl that have the same function as the eggshell? Explain.
3. **Key Concept** What does the structure of the eggshell tell you about its function?

Cell Shape and Movement

You might recall from Lesson 1 that all living things are made up of one or more cells. As illustrated in **Figure 4,** cells come in many shapes and sizes. The size and shape of a cell relates to its job or function. For example, a human red blood cell cannot be seen without a microscope. Its small size and disk shape enable it to pass easily through the smallest blood vessels. The shape of a nerve cell enables it to send signals over long distances. Some plant cells are hollow and make up tubelike structures that carry materials throughout a plant.

The structures that make up a cell also have unique functions. Think about how the players on a football team perform different tasks to move the ball down the field. In a similar way, a cell is made of different structures that perform different functions that keep a cell alive. You will read about some of these structures in this lesson.

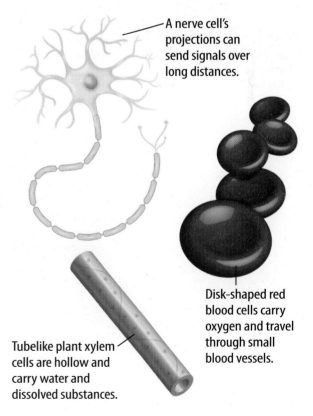

A nerve cell's projections can send signals over long distances.

Disk-shaped red blood cells carry oxygen and travel through small blood vessels.

Tubelike plant xylem cells are hollow and carry water and dissolved substances.

Figure 4 The shape of a cell relates to the function it performs.

Plant Cell

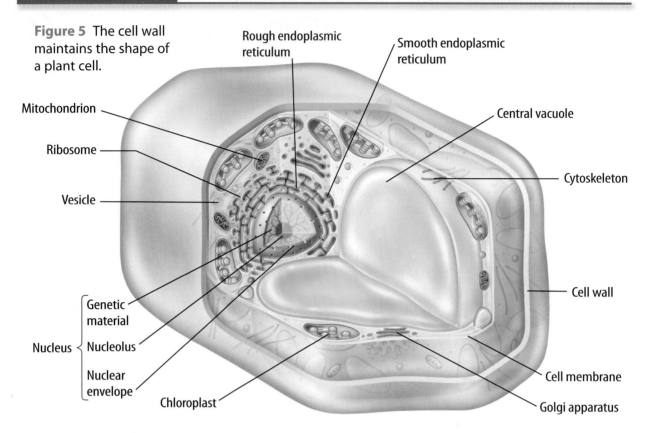

Figure 5 The cell wall maintains the shape of a plant cell.

Cell Membrane

Although different types of cells perform different *functions,* all cells have some structures in common. As shown in **Figure 5** and **Figure 6,** every cell is surrounded by a protective covering called a membrane. *The **cell membrane** is a flexible covering that protects the inside of a cell from the environment outside a cell.* Cell membranes are mostly made of two different macromolecules—proteins and a type of lipid called phospholipids. Think again about a football team. The defensive line tries to stop the other team from moving forward with the football. In a similar way, a cell membrane protects the cell from the outside environment.

 Reading Check What are cell membranes made of?

Cell Wall

Every cell has a cell membrane, but some cells are also surrounded by a structure called the cell wall. Plant cells such as the one in **Figure 5,** fungal cells, bacteria, and some types of protists have cell walls. *A **cell wall** is a stiff structure outside the cell membrane.* A cell wall protects a cell from attack by viruses and other harmful organisms. In some plant cells and fungal cells, a cell wall helps maintain the cell's shape and gives structural support.

ACADEMIC VOCABULARY
function
(noun) the purpose for which something is used

Animal Cell

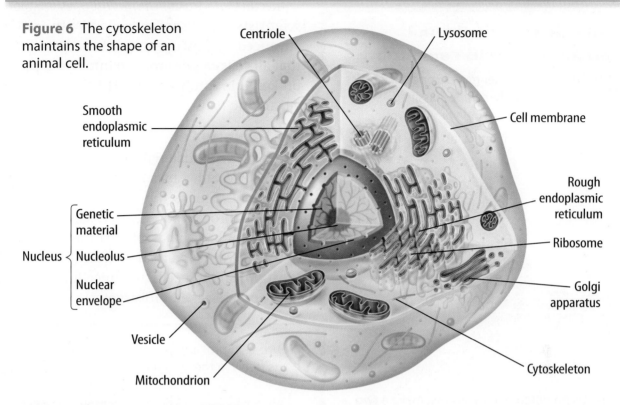

Figure 6 The cytoskeleton maintains the shape of an animal cell.

Visual Check Compare this animal cell to the plant cell in **Figure 5**.

Cell Appendages

Arms, legs, claws, and antennae are all types of appendages. Cells can have appendages too. Cell appendages are often used for movement. Flagella (fluh JEH luh; singular, flagellum) are long, tail-like appendages that whip back and forth and move a cell. A cell can also have cilia (SIH lee uh; singular, cilium) like the ones shown in **Figure 7.** Cilia are short, hairlike structures. They can move a cell or move molecules away from a cell. A microscopic organism called a paramecium (pa ruh MEE shee um) moves around its watery environment using its cilia. The cilia in your windpipe move harmful substances away from your lungs.

Figure 7 Lung cells have cilia that help move fluids and foreign materials.

Cytoplasm and the Cytoskeleton

In Lesson 1, you read that water is the main ingredient in a cell. Most of this water is in the **cytoplasm,** *a fluid inside a cell that contains salts and other molecules.* The cytoplasm also contains a cell's cytoskeleton. *The* **cytoskeleton** *is a network of threadlike proteins that are joined together.* The proteins form a framework inside a cell. This framework gives a cell its shape and helps it move. Cilia and flagella are made from the same proteins that make up the cytoskeleton.

WORD ORIGIN

cytoplasm
from Greek *kytos*, means "hollow vessel"; and *plasma,* means "something molded"

Inquiry MiniLab 25 minutes

How do eukaryotic and prokaryotic cells compare?

With the use of better microscopes, scientists discovered that cells can be classified as one of two types—prokaryotic or eukaryotic.

1. Read and complete a lab safety form.

2. Using different **craft items,** make a two-dimensional model of a eukaryotic cell.
3. In your cell model, include the number of cell structures assigned by your teacher.
4. Make each cell structure the correct shape, as shown in this lesson.
5. Make a label for each cell structure of your model.

Analyze and Conclude

1. **Describe** the nucleus of your cell.
2. **Classify** your cell as either a plant cell or an animal cell, and support your classification with evidence.
3. **Key Concept** Compare and contrast a prokaryotic cell, as shown in **Figure 8,** with your eukaryotic cell model.

Cell Types

Recall that the use of microscopes enabled scientists to discover cells. With more advanced microscopes, scientists discovered that all cells can be grouped into two types—prokaryotic (proh ka ree AH tihk) cells and eukaryotic (yew ker ee AH tihk) cells.

Prokaryotic Cells

The genetic material in a prokaryotic cell is not surrounded by a membrane, as shown in **Figure 8.** This is the most important feature of a prokaryotic cell. Prokaryotic cells also do not have many of the other cell parts that you will read about later in this lesson. Most prokaryotic cells are unicellular organisms and are called prokaryotes.

Figure 8 In prokaryotic cells, the genetic material floats freely in the cytoplasm.

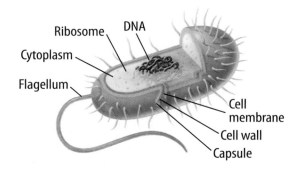

Eukaryotic Cells

Plants, animals, fungi, and protists are all made of eukaryotic cells, such as the ones shown in **Figure 5** and **Figure 6,** and are called eukaryotes. With few exceptions, each eukaryotic cell has genetic material that is surrounded by a membrane. Every eukaryotic cell also has *other structures, called* **organelles,** *which have specialized functions. Most organelles are surrounded by membranes.* Eukaryotic cells are usually larger than prokaryotic cells. About ten prokaryotic cells would fit inside one eukaryotic cell.

Key Concept Check How are prokaryotic cells and eukaryotic cells similar, and how are they different?

Cell Organelles

As you have just read, organelles are eukaryotic cell structures with specific functions. Organelles enable cells to carry out different functions at the same time. For example, cells can obtain energy from food, store information, make macromolecules, and get rid of waste materials all at the same time because different organelles perform the different tasks.

The Nucleus

The largest organelle inside most eukaryotic cells is the nucleus, shown in **Figure 9**. *The **nucleus** is the part of a eukaryotic cell that directs cell activities and contains genetic information stored in DNA.* DNA is organized into structures called chromosomes. The number of chromosomes in a nucleus is different for different species of organisms. For example, kangaroo cells contain six pairs of chromosomes. Most human cells contain 23 pairs of chromosomes.

Fold a sheet of paper into a vertical half book. Use it to record information about cell organelles and their functions.

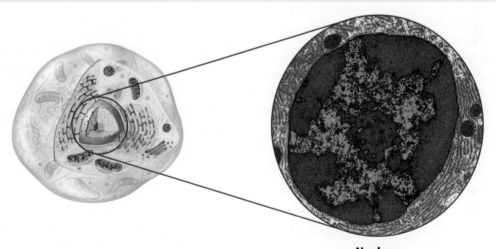

Nucleus
Color-Enhanced TEM Magnification: 15,500×

Figure 9 The nucleus directs cell activity and is surrounded by a membrane.

In addition to chromosomes, the nucleus contains proteins and an organelle called the nucleolus (new KLEE uh lus). The nucleolus is often seen as a large dark spot in the nucleus of a cell. The nucleolus makes ribosomes, organelles that are involved in the production of proteins. You will read about ribosomes later in this lesson.

Surrounding the nucleus are two membranes that form a structure called the nuclear envelope. The nuclear envelope contains many pores. Certain molecules, such as ribosomes and RNA, move into and out of the nucleus through these pores.

SCIENCE USE V. COMMON USE

envelope
Science Use an outer covering

Common Use a flat paper container for a letter

 Reading Check What is the nuclear envelope?

Lesson 2
EXPLAIN

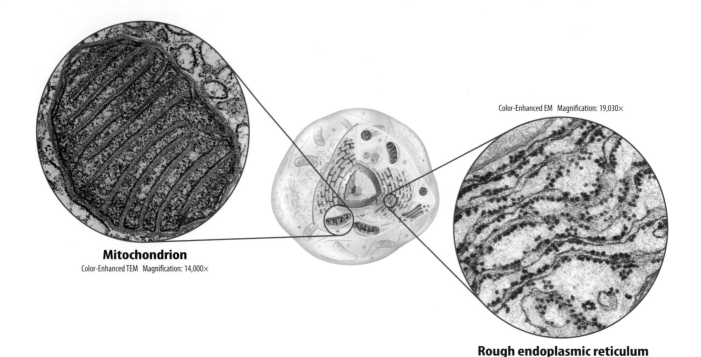

Mitochondrion
Color-Enhanced TEM Magnification: 14,000×

Color-Enhanced EM Magnification: 19,030×

Rough endoplasmic reticulum

Figure 10 The endoplasmic reticulum is made of many folded membranes. Mitochondria provide a cell with usable energy.

Manufacturing Molecules

You might recall from Lesson 1 that proteins are important molecules in cells. Proteins are made on small structures called ribosomes. Unlike other cell organelles, a ribosome is not surrounded by a membrane. Ribosomes are in a cell's cytoplasm. They also can be attached to a weblike organelle called the endoplasmic reticulum (en duh PLAZ mihk • rih TIHK yuh lum), or ER. As shown in **Figure 10,** the ER spreads from the nucleus throughout most of the cytoplasm. ER with ribosomes on its surface is called rough ER. Rough ER is the site of protein production. ER without ribosomes is called smooth ER. It makes lipids such as cholesterol. Smooth ER is important because it helps remove harmful substances from a cell.

 Reading Check Contrast smooth ER and rough ER.

Processing Energy

All living things require energy in order to survive. Cells process some energy in specialized organelles. Most eukaryotic cells contain hundreds of organelles called mitochondria (mi tuh KAHN dree uh; singular, mitochondrion), shown in **Figure 10.** Some cells in a human heart can contain a thousand mitochondria.

Like the nucleus, a mitochondrion is surrounded by two membranes. Energy is released during chemical reactions that occur in the mitochondria. This energy is stored in high-energy molecules called ATP—adenosine triphosphate (uh DEH nuh seen • tri FAHS fayt). ATP is the fuel for cellular processes such as growth, cell division, and material transport.

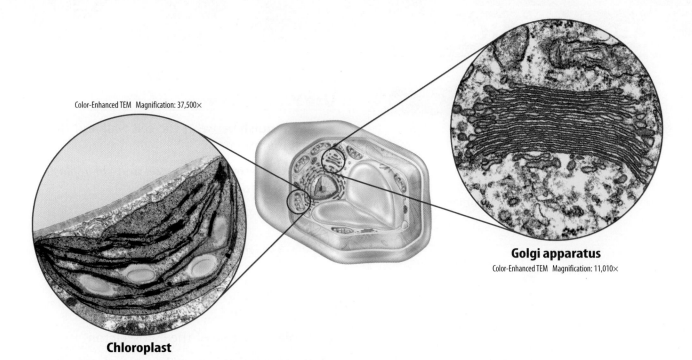

Chloroplast

Golgi apparatus
Color-Enhanced TEM Magnification: 11,010×

Color-Enhanced TEM Magnification: 37,500×

Plant cells and some protists, such as algae, also contain organelles called chloroplasts (KLOR uh plasts), shown in **Figure 11**. **Chloroplasts** *are membrane-bound organelles that use light energy and make food—a sugar called glucose—from water and carbon dioxide in a process known as photosynthesis* (foh toh SIHN thuh sus). The sugar contains stored chemical energy that can be released when a cell needs it. You will read more about photosynthesis in Lesson 4.

Figure 11 Plant cells have chloroplasts that use light energy and make food. The Golgi apparatus packages materials into vesicles.

 Reading Check Which types of cells contain chloroplasts?

Processing, Transporting, and Storing Molecules

Near the ER is an organelle that looks like a stack of pancakes. This is the Golgi (GAWL jee) apparatus, shown in **Figure 11**. It prepares proteins for their specific jobs or functions. Then it packages the proteins into tiny, membrane-bound, ball-like structures called vesicles. Vesicles are organelles that transport substances from one area of a cell to another area of a cell. Some vesicles in an animal cell are called lysosomes. Lysosomes contain substances that help break down and recycle cellular components.

Some cells also have saclike structures called vacuoles (VA kyuh wohlz). Vacuoles are organelles that store food, water, and waste material. A typical plant cell usually has one large vacuole that stores water and other substances. Some animal cells have many small vacuoles.

 Key Concept Check What is the function of the Golgi apparatus?

Lesson 2 Review

Visual Summary

A cell is protected by a flexible covering called the cell membrane.

Cells can be grouped into two types—prokaryotic cells and eukaryotic cells.

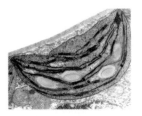
In a chloroplast, light energy is used for making sugars in a process called photosynthesis.

FOLDABLES

Use your lesson Foldable to review the lesson. Save your Foldable for the project at the end of the chapter.

What do you think NOW?

You first read the statements below at the beginning of the chapter.

3. Different organisms have cells with different structures.

4. All cells store genetic information in their nuclei.

Did you change your mind about whether you agree or disagree with the statements? Rewrite any false statements to make them true.

Use Vocabulary

1. **Distinguish** between the cell wall and the cell membrane.

2. **Use the terms** *mitochondria* and *chloroplasts* in a sentence.

3. **Define** *organelle* in your own words.

Understand Key Concepts

4. Which organelle is used to store water?
   A. chloroplast   C. nucleus
   B. lysosome   D. vacuole

5. **Explain** the role of the cytoskeleton.

6. **Draw** a prokaryotic cell and label its parts.

7. **Compare** the roles of the endoplasmic reticulum and the Golgi apparatus.

Interpret Graphics

8. **Explain** how the structure of the cells below relates to their function.

9. **Compare** Copy the table below and fill it in to compare the structures of a plant cell to the structures of an animal cell.

Structure	Plant Cell	Animal Cell
Cell membrane	yes	yes
Cell wall		
Mitochondrion		
Chloroplast		
Nucleus		
Vacuole		
Lysosome		

Critical Thinking

10. **Analyze** Why are most organelles surrounded by membranes?

11. **Compare** the features of eukaryotic and prokaryotic cells.

Inquiry Skill Practice: Compare and Contrast

45 minutes

How are plant cells and animal cells similar and how are they different?

A light microscope enables you to observe many of the structures in cells. Increasing the magnification means you see a smaller portion of the object, but lets you see more detail. As you see more details, you can **compare and contrast** different cell types. How are they alike? How are they different?

Materials

microscope

microscope slide and coverslip

forceps

dropper

Elodea plant

Prepared slide of human cheek cells

Safety

Learn It

Observations can be analyzed by noting the similarities and differences between two or more objects that you observe. You **compare** objects by noting similarities. You **contrast** objects by looking for differences.

Try It

1. Read and complete a lab safety form.

2. Using forceps, make a wet-mount slide of a young leaf from the tip of an *Elodea* plant.

3. Use a microscope to observe the leaf on low power. Focus on the top layer of cells.

4. Switch to high power and focus on one cell. The large organelle in the center of the cell is the central vacuole. Moving around the central vacuole are green, disklike objects called chloroplasts. Try to find the nucleus. It looks like a clear ball.

5. Draw a diagram of an *Elodea* cell in your Science Journal. Label the cell wall, central vacuole, chloroplasts, cytoplasm, and nucleus. Return to low power and remove the slide. Properly dispose of the slide.

6. Observe the prepared slide of cheek cells under low power.

7. Switch to high power and focus on one cell. Draw a diagram of one cheek cell. Label the cell membrane, cytoplasm, and nucleus. Return to low power and remove the slide.

Apply It

8. Based on your diagrams, how do the shapes of the *Elodea* cell and cheek cell compare?

9. **Key Concept** Compare and contrast the cell structures in your two diagrams. Which structures did you observe in both cells? Which structures did you observe in only one of the cells?

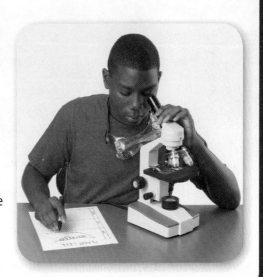

Lesson 2 • 59
EXTEND

Lesson 3

Reading Guide

Key Concepts 🔑
ESSENTIAL QUESTIONS

- How do materials enter and leave cells?
- How does cell size affect the transport of materials?

Vocabulary

passive transport p. 61
diffusion p. 62
osmosis p. 62
facilitated diffusion p. 63
active transport p. 64
endocytosis p. 64
exocytosis p. 64

g Multilingual eGlossary

Moving Cellular Material

Inquiry Why the Veil?

A beekeeper often wears a helmet with a face-covering veil made of mesh. The openings in the mesh are large enough to let air through, yet small enough to keep bees out. In a similar way, some things must be allowed in or out of a cell, while other things must be kept in or out. How do the right things enter or leave a cell?

Inquiry Launch Lab

5 minutes

What does the cell membrane do?

All cells have a membrane around the outside of the cell. The cell membrane separates the inside of a cell from the environment outside a cell. What else might a cell membrane do?

1. Read and complete a lab safety form.
2. Place a square of **wire mesh** on top of a **beaker**.
3. Pour a small amount of **birdseed** on top of the wire mesh. Record your observations in your Science Journal.

Think About This

1. What part of a cell does the wire mesh represent?
2. What happened when you poured birdseed on the wire mesh?
3. **Key Concept** How do you think the cell membrane affects materials that enter and leave a cell?

Passive Transport

Recall from Lesson 2 that membranes are the boundaries between cells and between organelles. Another important role of membranes is to control the movement of substances into and out of cells. A cell membrane is semipermeable. This means it allows only certain substances to enter or leave a cell. Substances can pass through a cell membrane by one of several different processes. The type of process depends on the physical and chemical properties of the substance passing through the membrane.

Small molecules, such as oxygen and carbon dioxide, pass through membranes by a process called passive transport. **Passive transport** *is the movement of substances through a cell membrane without using the cell's energy.* Passive transport depends on the amount of a substance on each side of a membrane. For example, suppose there are more molecules of oxygen outside a cell than inside it. Oxygen will move into that cell until the amount of oxygen is equal on both sides of the cell's membrane. Since oxygen is a small molecule, it passes through a cell membrane without using the cell's energy. The different types of passive transport are explained on the following pages.

Reading Check Describe a semipermeable membrane.

FOLDABLES

Fold a sheet of paper into a two-tab book. Label the tabs as shown. Use it to organize information about the different types of passive and active transport.

Diffusion

What happens when the concentration, or amount per unit of volume, of a substance is unequal on each side of a membrane? The molecules will move from the side with a higher concentration of that substance to the side with a lower concentration. **Diffusion** *is the movement of substances from an area of higher concentration to an area of lower concentration.*

Usually, diffusion continues through a membrane until the concentration of a substance is the same on both sides of the membrane. When this happens, a substance is in equilibrium. Compare the two diagrams in **Figure 12.** What happened to the red dye that was added to the water on one side of the membrane? Water and dye passed through the membrane in both directions until there were equal concentrations of water and dye on both sides of the membrane.

WORD ORIGIN
diffusion
from Latin *diffusionem*, means "scatter, pour out"

Diffusion

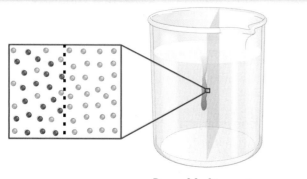
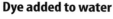
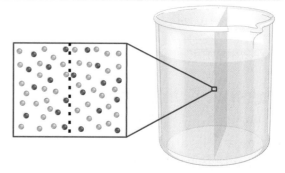

Dye added to water   **After 30 minutes**

✓ **Visual Check** What would the water in the beaker on the right look like if the membrane did not let anything through?

Figure 12 Over time, the concentration of dye on either side of the membrane becomes the same.

Osmosis—The Diffusion of Water

Diffusion refers to the movement of any small molecules from higher to lower concentrations. However, **osmosis** *is the diffusion of water molecules only through a membrane.* Semipermeable cell membranes also allow water to pass through them until equilibrium occurs. For example, the amount of water stored in the vacuoles of plant cells can decrease because of osmosis. That is because the concentration of water in the air surrounding the plant is less than the concentration of water inside the vacuoles of plant cells. Water will continue to diffuse into the air until the concentrations of water inside the plant's cells and in the air are equal. If the plant is not watered to replace the lost water, it will wilt and eventually die.

Facilitated Diffusion

Some molecules are too large or are chemically unable to travel through a membrane by diffusion. *When molecules pass through a cell membrane using special proteins called transport proteins, this is* **facilitated diffusion.** Like diffusion and osmosis, facilitated diffusion does not require a cell to use energy. As shown in **Figure 13,** a cell membrane has transport proteins. The two types of transport proteins are carrier proteins and channel proteins. Carrier proteins carry large molecules, such as the sugar molecule glucose, through the cell membrane. Channel proteins form pores through the membrane. Atomic particles, such as sodium ions and potassium ions, pass through the cell membrane by channel proteins.

 Reading Check How do materials move through the cell membrane in facilitated diffusion?

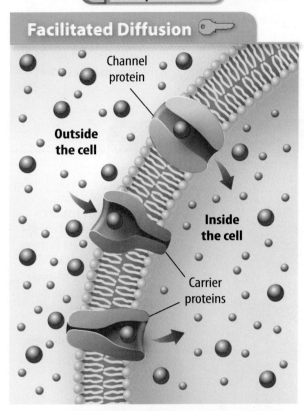

Figure 13 Transport proteins are used to move large molecules into and out of a cell.

Inquiry MiniLab   20 minutes

How is a balloon like a cell membrane?

Substances within a cell are constantly in motion. How can a balloon act like a cell membrane?

1. Read and complete a lab safety form.
2. Make a three-column table in your Science Journal to record your data. Label the first column *Balloon Number*, the second column *Substance*, and the third column *Supporting Evidence.*
3. Use your senses to identify what substance is in each of the **numbered balloons.**
4. Record what you think each substance is.
5. Record the evidence supporting your choice.

Analyze and Conclude

1. **List** the senses that were most useful in identifying the substances.
2. **Infer** if you could identify the substances if you were blindfolded. If so, how?
3. **Describe** how the substances moved, and explain why they moved this way.
4. **Key Concept** Explain how a balloon is like a cell membrane in terms of the movement of substances.

Lesson 3
EXPLAIN

Figure 14 Active transport is most often used to bring needed nutrients into a cell. Endocytosis and exocytosis move materials that are too large to pass through the cell membrane by other methods.

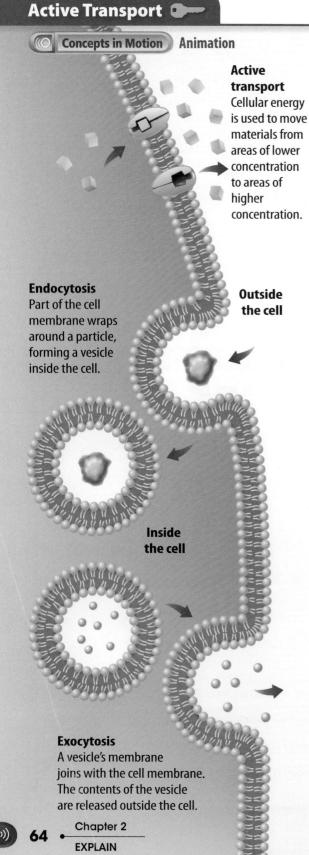

Active Transport

Sometimes when cellular materials pass through membranes it requires a cell to use energy. **Active transport** *is the movement of substances through a cell membrane only by using the cell's energy.*

Recall that passive transport is the movement of substances from areas of higher concentration to areas of lower concentration. However, substances moving by active transport move from areas of lower concentration to areas of higher concentration, as shown in **Figure 14**.

Active transport is important for cells and organelles. Cells can take in needed nutrients from the environment through carrier proteins by using active transport. This occurs even when concentrations of these nutrients are lower in the environment than inside the cell. Some other molecules and waste materials also leave cells by active transport.

Endocytosis and Exocytosis

Some substances are too large to enter a cell membrane by diffusion or by using a transport protein. These substances can enter a cell by another process. **Endocytosis** (en duh si TOH sus), shown in **Figure 14**, *is the process during which a cell takes in a substance by surrounding it with the cell membrane.* Many different types of cells use endocytosis. For example, some cells take in bacteria and viruses using endocytosis.

Some substances are too large to leave a cell by diffusion or by using a transport protein. These substances can leave a cell another way. **Exocytosis** (ek soh si TOH sus), shown in **Figure 14**, *is the process during which a cell's vesicles release their contents outside the cell.* Proteins and other substances are removed from a cell through this process.

 Key Concept Check How do materials enter and leave cells?

Cell Size and Transport

Recall that the movement of nutrients, waste material, and other substances into and out of a cell is important for survival. For this movement to happen, the area of the cell membrane must be large compared to its volume. The area of the cell membrane is the cell's surface area. The volume is the amount of space inside the cell. As a cell grows, both its volume and its surface area increase. The volume of a cell increases faster than its surface area. If a cell were to keep growing, it would need large amounts of nutrients and would produce large amounts of waste material. However, the surface area of the cell's membrane would be too small to move enough nutrients and wastes through it for the cell to survive.

Key Concept Check How does cell size affect the transport of materials?

Math Skills — Use Ratios

Review
- Math Practice
- Personal Tutor

A ratio is a comparison of two numbers, such as surface area and volume. If a cell were cube-shaped, you would calculate surface area by multiplying its length (ℓ) by its width (w) by the number of sides (6).

Surface area $= \ell \times w \times 6$

You would calculate the volume of the cell by multiplying its length (ℓ) by its width (w) by its height (h).

Volume $= \ell \times w \times h$

To find the surface-area-to-volume ratio of the cell, divide its surface area by its volume.

$$\frac{\text{Surface area}}{\text{Volume}}$$

In the table below, surface-area-to-volume ratios are calculated for cells that are 1 mm, 2 mm, and 4 mm per side. Notice how the ratios change as the cell's size increases.

Length	1 mm	2 mm	4 mm
Width	1 mm	2 mm	4 mm
Height	1 mm	2 mm	4 mm
Number of sides	6	6	6
Surface area ($\ell \times w \times$ no. of sides)	1 mm × 1 mm × 6 = 6 mm²	2 mm × 2 mm × 6 = 24 mm²	4 mm × 4 mm × 6 = 96 mm²
Volume ($\ell \times w \times h$)	1 mm × 1 mm × 1 mm = 1 mm³	2 mm × 2 mm × 2 mm = 8 mm³	4 mm × 4 mm × 4 mm = 64 mm³
Surface-area-to-volume ratio	$\frac{6 \text{ mm}^2}{1 \text{ mm}^3} = \frac{6}{1}$ or 6:1	$\frac{24 \text{ mm}^2}{8 \text{ mm}^3} = \frac{3}{1}$ or 3:1	$\frac{96 \text{ mm}^2}{64 \text{ mm}^3} = \frac{1.5}{1}$ or 1.5:1

Practice

What is the surface-area-to-volume ratio of a cell whose six sides are 3 mm long?

Lesson 3 Review

Visual Summary

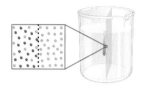

Small molecules can move from an area of higher concentration to an area of lower concentration by diffusion.

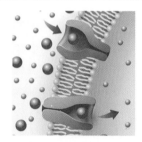

In facilitated diffusion, proteins transport larger molecules through a cell membrane.

Some molecules move from areas of lower concentration to areas of higher concentration through active transport.

FOLDABLES

Use your lesson Foldable to review the lesson. Save your Foldable for the project at the end of the chapter.

What do you think NOW?

You first read the statements below at the beginning of the chapter.

5. Diffusion and osmosis are the same process.

6. Cells with large surface areas can transport more than cells with smaller surface areas.

Did you change your mind about whether you agree or disagree with the statements? Rewrite any false statements to make them true.

Use Vocabulary

1 **Use the term** *osmosis* in a sentence.

2 **Distinguish** between active transport and passive transport.

3 The process by which vesicles move substances out of a cell is _____.

Understand Key Concepts

4 **Explain** why energy is needed in active transport.

5 **Summarize** the function of endocytosis.

6 **Contrast** osmosis and diffusion.

7 What is limited by a cell's surface-area-to-volume ratio?
  A. cell shape   C. cell surface area
  B. cell size   D. cell volume

Interpret Graphics

8 **Identify** the process shown below, and explain how it works.

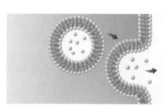

9 **Copy** and fill in the graphic organizer below to describe ways that cells transport substances.

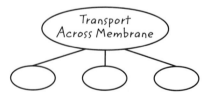

Critical Thinking

10 **Relate** the surface area of a cell to the transport of materials.

Math Skills

— Math Practice —

11 **Calculate** the surface-area-to-volume ratio of a cube whose sides are 6 cm long.

Inquiry Skill Practice — Analyze and Conclude 30 minutes

How does an object's size affect the transport of materials?

Materials

hard-cooked eggs

metric ruler

blue food coloring

250-mL beaker

plastic spoon

plastic knife

paper towels

Safety

Nutrients, oxygen, and other materials enter and leave a cell through the cell membrane. Does the size of a cell affect the transport of these materials throughout the cell? In this lab, you will **analyze and conclude** how the size of a cube of egg white affects material transport.

Learn It

To **analyze** how an object's size affects material transport, you will need to calculate each object's surface-area-to-volume ratio. The following formulas are used to calculate surface area and volume of a cube.

surface area (mm^2) = (length of 1 side)^2 × 6

volume (mm^3) = (length of 1 side)^3

To calculate the ratio of surface area to volume, divide surface area by volume.

Try It

1. Read and complete a lab safety form.

2. Measure and cut one large cube of egg white that is 20 mm on each side. Then, measure and cut one small cube of egg white that is 10 mm on each side.

3. Place 100 mL of water in a plastic cup. Add 10 drops of food coloring. Gently add the egg-white cubes, and soak overnight.

4. Remove the cubes from the cup with a plastic spoon and place them on a paper towel. Cut each cube in half.

5. Examine the inside surface of each cube. Measure and record in millimeters how deep the blue food coloring penetrated into each cube.

Apply It

6. How does the depth of the color compare on the two cubes?

7. Calculate the surface area, the volume, and the surface-area-to-volume ratio of each cube. How do the surface-area-to-volume ratios of the two cubes compare?

8. 🔑 **Key Concept** Would a cell with a small surface-area-to-volume ratio be able to transport nutrients and waste through the cell as efficiently as a cell with a large surface-area-to-volume ratio?

Lesson 3
EXTEND
67

Lesson 4

Cells and Energy

Reading Guide

Key Concepts
ESSENTIAL QUESTIONS

- How does a cell obtain energy?
- How do some cells make food molecules?

Vocabulary

cellular respiration p. 69
glycolysis p. 69
fermentation p. 70
photosynthesis p. 71

 Multilingual eGlossary

Inquiry Why are there bubbles?

Have you ever seen bubbles on a green plant in an aquarium? Where did the bubbles come from? Green plants use light energy and make sugars and oxygen.

68 Chapter 2
ENGAGE

Inquiry Launch Lab

5 minutes

What do you exhale?
Does the air you breathe in differ from the air you breathe out?

1. Read and complete a lab safety form.
2. Unwrap a **straw.** Use the straw to slowly blow into a small **cup** of **bromthymol blue.** Do not splash the liquid out of the cup.
3. In your Science Journal, record any changes in the solution.

Think About This
1. What changes did you observe in the solution?
2. What do you think caused the changes in the solution?
3. **Key Concept** Why do you think the air you inhale differs from the air you exhale?

Cellular Respiration

When you are tired, you might eat something to give you energy. All living things, from one-celled organisms to humans, need energy to survive. Recall that cells process energy from food into the energy-storage compound ATP. **Cellular respiration** is a *series of chemical reactions that convert the energy in food molecules into a usable form of energy called ATP.* Cellular respiration is a complex process that occurs in two parts of a cell—the cytoplasm and the mitochondria.

Reactions in the Cytoplasm

The first step of cellular respiration, called glycolysis, occurs in the cytoplasm of all cells. **Glycolysis** is *a process by which glucose, a sugar, is broken down into smaller molecules.* As shown in **Figure 15,** glycolysis produces some ATP molecules. It also uses energy from other ATP molecules. You will read on the following page that more ATP is made during the second step of cellular respiration than during glycolysis.

Reading Check What is produced during glycolysis?

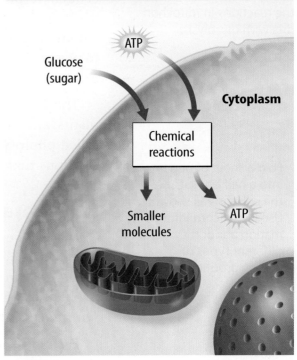

Glycolysis

Figure 15 Glycolysis is the first step of cellular respiration.

Reactions in the Mitochondria

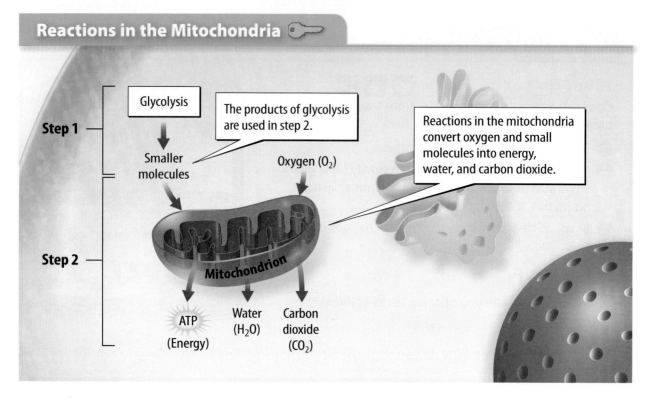

Figure 16 After glycolysis, cellular respiration continues in the mitochondria.

Visual Check Compare the reactions in mitochondria with glycolysis.

FOLDABLES
Fold a sheet of paper into a half book. Label the columns as shown. Use it to record information about the different types of energy production.

Reactions in the Mitochondria

The second step of cellular respiration occurs in the mitochondria of eukaryotic cells, as shown in **Figure 16.** This step of cellular respiration requires oxygen. The smaller molecules made from glucose during glycolysis are broken down. Large amounts of ATP—usable energy—are produced. Cells use ATP to power all cellular processes. Two waste products—water and carbon dioxide (CO_2)—are given off during this step.

The CO_2 released by cells as a waste product is used by plants and some unicellular organisms in another process called photosynthesis. You will read more about the chemical reactions that take place during photosynthesis in this lesson.

Fermentation

Have you ever felt out of breath after exercising? Sometimes when you exercise, your cells don't have enough oxygen to make ATP through cellular respiration. Then, chemical energy is obtained through a different process called fermentation. This process does not use oxygen.

Fermentation *is a reaction that eukaryotic and prokaryotic cells can use to obtain energy from food when oxygen levels are low.* Because no oxygen is used, fermentation makes less ATP than cellular respiration does. Fermentation occurs in a cell's cytoplasm, not in mitochondria.

Key Concept Check How does a cell obtain energy?

Types of Fermentation

One type of fermentation occurs when glucose is converted into ATP and a waste product called lactic acid, as illustrated in **Figure 17**. Some bacteria and fungi help produce cheese, yogurt, and sour cream using lactic-acid fermentation. Muscle cells in humans and other animals can use lactic-acid fermentation and obtain energy during exercise.

Some types of bacteria and yeast make ATP through a process called alcohol fermentation. However, instead of producing lactic acid, alcohol fermentation produces an alcohol called ethanol and CO_2, also illustrated in **Figure 17**. Some types of breads are made using yeast. The CO_2 produced by yeast during alcohol fermentation makes the dough rise.

 Reading Check Compare lactic-acid fermentation and alcohol fermentation.

Figure 17 Your muscle cells produce lactic acid as a waste during fermentation. Yeast cells produce carbon dioxide and alcohol as wastes during fermentation.

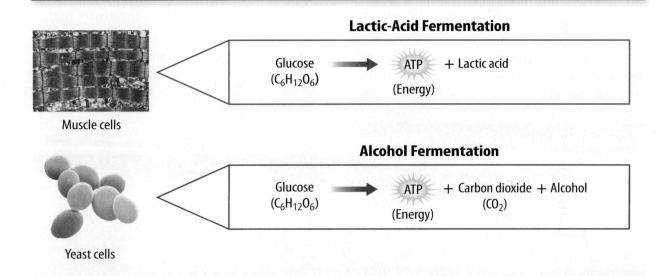

Photosynthesis

Humans and other animals convert food energy into ATP through cellular respiration. However, plants and some unicellular organisms obtain energy from light. **Photosynthesis** *is a series of chemical reactions that convert light energy, water, and CO_2 into the food-energy molecule glucose and give off oxygen.*

Lights and Pigments

Photosynthesis requires light energy. In plants, pigments such as chlorophyll absorb light energy. When chlorophyll absorbs light, it absorbs all colors except green. Green light is reflected as the green color seen in leaves. However, plants contain many pigments that reflect other colors, such as yellow and red.

WORD ORIGIN

photosynthesis
from Greek *photo*, means "light"; and *synthesis*, means "composition"

Reactions in Chloroplasts

The light energy absorbed by chlorophyll and other pigments powers the chemical reactions of photosynthesis. These reactions occur in chloroplasts, the organelles in plant cells that convert light energy to chemical energy in food. During photosynthesis, light energy, water, and carbon dioxide combine and make sugars. Photosynthesis also produces oxygen that is released into the atmosphere, as shown in **Figure 18.**

 Key Concept Check How do some cells make food molecules?

Importance of Photosynthesis

Recall that photosynthesis uses light energy and CO_2 and makes food energy and releases oxygen. This food energy is stored in the form of glucose. When an organism, such as the bird in **Figure 18,** eats plant material, such as fruit, it takes in food energy. An organism's cells use the oxygen released during photosynthesis and convert the food energy into usable energy through cellular respiration. **Figure 18** illustrates the important relationship between cellular respiration and photosynthesis.

Figure 18 The relationship between cellular respiration and photosynthesis is important for life.

Lesson 4 Review

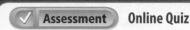

Visual Summary

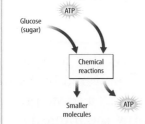

Glycolysis is the first step in cellular respiration.

Fermentation provides cells, such as muscle cells, with energy when oxygen levels are low.

Light energy powers the chemical reactions of photosynthesis.

FOLDABLES

Use your lesson Foldable to review the lesson. Save your Foldable for the project at the end of the chapter.

What do you think NOW?

You first read the statements below at the beginning of the chapter.

7. ATP is the only form of energy found in cells.

8. Cellular respiration occurs only in lung cells.

Did you change your mind about whether you agree or disagree with the statements? Rewrite any false statements to make them true.

Use Vocabulary

1 **Define** *glycolysis* using your own words.

2 **Distinguish** between cellular respiration and fermentation.

3 A process used by plants to convert light energy into food energy is _____.

Understand Key Concepts

4 Which contains pigments that absorb light energy?
  A. chloroplast   C. nucleus
  B. mitochondrion   D. vacuole

5 **Relate** mitochondria to cellular respiration.

6 **Describe** the role of chlorophyll in photosynthesis.

7 **Give an example** of how fermentation is used in the food industry.

Interpret Graphics

8 **Draw** a graphic organizer like the one below. Fill in the boxes with the substances used and produced during photosynthesis.

9 **Summarize** the steps of cellular respiration using the figure below.

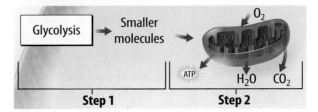

Critical Thinking

10 **Design** a concept map to show the relationship between cellular respiration in animals and photosynthesis in plants.

11 **Summarize** the roles of glucose and ATP in energy processing.

Lesson 4
EVALUATE

Inquiry Lab

50 minutes

Materials

test tube

Elodea

scissors

beaker

lamp

watch or clock

thermometer

Safety

Photosynthesis and Light

You might think of photosynthesis as a process of give and take. Plant cells take in water and carbon dioxide, and, powered by light energy, make their own food. Plants give off oxygen as a waste product during photosynthesis. Can you determine how the intensity of light affects the rate of photosynthesis?

Ask a Question

How does the intensity of light affect photosynthesis?

Make Observations

1. Read and complete a lab safety form.

2. Cut the bottom end of an *Elodea* stem at an angle, and lightly crush the cut end. Place the *Elodea* in a test tube with the cut side at the top. Fill the test tube with water. Stand the test tube and a thermometer in a beaker filled with water. (The water in the beaker keeps the water in the test tube from getting too warm under the lamp.)

3. Place the beaker containing your test tube on a sheet of paper under a lamp. Measure the temperature of the water in the beaker. Record the temperature in your Science Journal.

4. When bubbles of oxygen begin to rise from the plant, start counting the number of bubbles per minute. Continue to record this data for 10 minutes.

5. Record the temperature of the water in the beaker at the end of the test.

6. Calculate the average number of bubbles produced per minute by your plant.

7. Compare your data with your classmates' data.

Form a Hypothesis

8 Use your data to form a hypothesis relating the amount of light to the rate of photosynthesis.

Test Your Hypothesis

9 Repeat the experiment, changing the light variable so that you are observing your plant's reaction to getting either more or less light. An increase or decrease in water temperature will indicate a change in the amount of light. Keep all other conditions the same.

10 Record your data in a table similar to the one shown at right, and calculate the average number of bubbles per minute.

Analyze and Conclude

11 Use Variables How does the amount of light affect photosynthesis? What is your evidence?

12 The Big Idea How do plant cells make food? What do they take in and what do they give off? What source of energy do they use?

Communicate Your Results

Compile all the class data on one graph to show the effects of varying amounts of light on the rate of photosynthesis.

Number of Bubbles per Minute		
Time	Control	Less Light
1		
2		
3		
4		
5		
6		
7		
8		
9		
10		

Inquiry Extension

What other variables might affect the rate of photosynthesis? For example, how does different-colored light or a change in temperature affect the rate of photosynthesis? To investigate your question, design a controlled experiment.

Lab Tips

☑ To calculate the average number of bubbles per minute, add the total number of bubbles observed in 10 minutes, and then divide by 10.

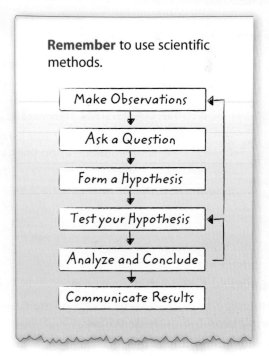

Remember to use scientific methods.

Make Observations → Ask a Question → Form a Hypothesis → Test your Hypothesis → Analyze and Conclude → Communicate Results

Chapter 2 Study Guide

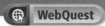

THE BIG IDEA: A cell is made up of structures that provide support and movement; process energy; and transport materials into, within, and out of a cell.

Key Concepts Summary

Lesson 1: Cells and Life

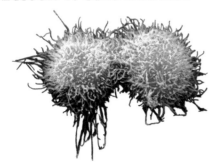

- The invention of the microscope led to discoveries about cells. In time, scientists used these discoveries to develop the **cell theory**, which explains how cells and living things are related.
- Cells are composed mainly of water, **proteins, nucleic acids, lipids,** and **carbohydrates.**

Lesson 2: The Cell

- Cell structures have specific functions, such as supporting a cell, moving a cell, controlling cell activities, processing energy, and transporting molecules.
- A prokaryotic cell lacks a nucleus and other **organelles,** while a eukaryotic cell has a nucleus and other organelles.

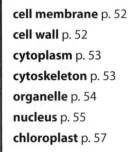

Lesson 3: Moving Cellular Material

- Materials enter and leave a cell through the cell membrane using **passive transport** or **active transport, endocytosis,** and **exocytosis.**
- The ratio of surface area to volume limits the size of a cell. In a smaller cell, the high surface-area-to-volume ratio allows materials to move easily to all parts of a cell.

Lesson 4: Cells and Energy

- All living cells release energy from food molecules through **cellular respiration** and/or **fermentation.**
- Some cells make food molecules using light energy through the process of **photosynthesis.**

$$C_6H_{12}O_6 + 6O_2 \longrightarrow 6CO_2 + 6H_2O + ATP \text{ (Energy)}$$

Cellular respiration

$$6CO_2 + 6H_2O \xrightarrow[\text{Chlorophyll}]{\text{Light energy}} C_6H_{12}O_6 + 6O_2$$

Photosynthesis

Vocabulary

Lesson 1:
- cell theory p. 44
- macromolecule p. 45
- nucleic acid p. 46
- protein p. 47
- lipid p. 47
- carbohydrate p. 47

Lesson 2:
- cell membrane p. 52
- cell wall p. 52
- cytoplasm p. 53
- cytoskeleton p. 53
- organelle p. 54
- nucleus p. 55
- chloroplast p. 57

Lesson 3:
- passive transport p. 61
- diffusion p. 62
- osmosis p. 62
- facilitated diffusion p. 63
- active transport p. 64
- endocytosis p. 64
- exocytosis p. 64

Lesson 4:
- cellular respiration p. 69
- glycolysis p. 69
- fermentation p. 70
- photosynthesis p. 71

Study Guide

- Vocabulary eFlashcards
- Vocabulary eGames
- Personal Tutor

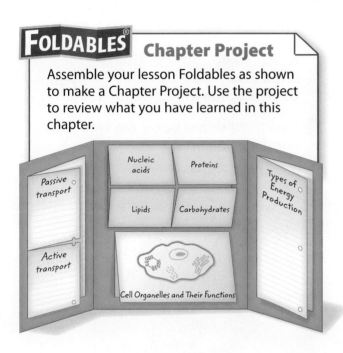

Chapter Project

Assemble your lesson Foldables as shown to make a Chapter Project. Use the project to review what you have learned in this chapter.

Use Vocabulary

1. Substances formed by joining smaller molecules together are called _____.

2. The _____ consists of proteins joined together to create fiberlike structures inside cells.

3. The movement of substances from an area of high concentration to an area of low concentration is called _____.

4. A process that uses oxygen to convert energy from food into ATP is _____ _____.

Link Vocabulary and Key Concepts

Interactive Concept Map

Copy this concept map, and then use vocabulary terms from the previous page to complete the concept map.

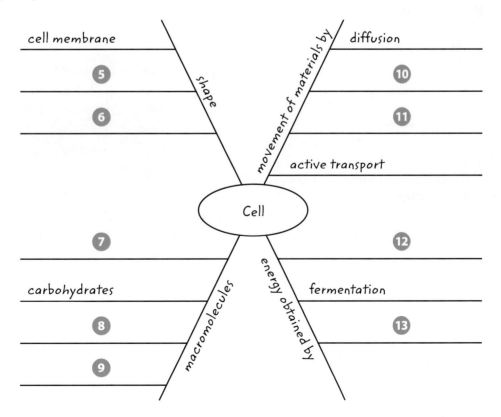

Chapter 2 Review

Understand Key Concepts

1 Cholesterol is which type of macromolecule?
  A. carbohydrate
  B. lipid
  C. nucleic acid
  D. protein

2 Genetic information is stored in which macromolecule?
  A. DNA
  B. glucose
  C. lipid
  D. starch

3 The arrow below is pointing to which cell part?

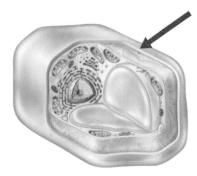

  A. chloroplast
  B. mitochondrion
  C. cell membrane
  D. cell wall

4 Which best describes vacuoles?
  A. lipids
  B. proteins
  C. contained in mitochondria
  D. storage compartments

5 Which is true of fermentation?
  A. does not generate energy
  B. does not require oxygen
  C. occurs in mitochondria
  D. produces lots of ATP

6 Which process eliminates substances from cells in vesicles?
  A. endocytosis
  B. exocytosis
  C. osmosis
  D. photosynthesis

7 Which cell shown below can send signals over long distances?

  A.
  B.
  C.
  D.

8 The figure below shows a cell. What is the arrow pointing to?

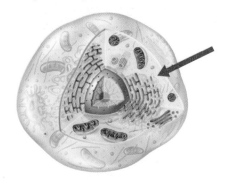

  A. chloroplast
  B. cytoplasm
  C. mitochondrion
  D. nucleus

Chapter Review

Assessment
Online Test Practice

Critical Thinking

9 **Evaluate** the importance of the microscope to biology.

10 **Summarize** the role of water in cells.

11 **Hypothesize** how new cells form from existing cells.

12 **Distinguish** between channel proteins and carrier proteins.

13 **Explain** osmosis.

14 **Infer** Why do cells need carrier proteins that transport glucose?

15 **Compare** the amounts of ATP generated in cellular respiration and fermentation.

16 **Assess** the role of fermentation in baking bread.

17 **Hypothesize** how air pollution like smog affects photosynthesis.

18 **Compare** prokaryotes and eukaryotes by copying and filling in the table below.

Structure	Prokaryote (yes or no)	Eukaryote (yes or no)
Cell membrane		
DNA		
Nucleus		
Endoplasmic reticulum		
Golgi apparatus		
Cell wall		

Writing in Science

19 **Write** a five-sentence paragraph relating the cytoskeleton to the walls of a building. Be sure to include a topic sentence and a concluding sentence in your paragraph.

REVIEW THE BIG IDEA

20 How do the structures and processes of a cell enable it to survive? As an example, explain how chloroplasts help plant cells.

21 The photo below shows a protozoan. What structures enable it to get food into its mouth?

Math Skills

Review — Math Practice

Use Ratios

22 A rectangular solid measures 4 cm long by 2 cm wide by 2 cm high. What is the surface-area-to-volume ratio of the solid?

23 At different times during its growth, a cell has the following surface areas and volumes:

Time	Surface area (μm)	Volume (μm)
1	6	1
2	24	8
3	54	27

What happens to the surface-area-to-volume ratio as the cell grows?

Chapter 2 Review • 79

Standardized Test Practice

Record your answers on the answer sheet provided by your teacher or on a sheet of paper.

Multiple Choice

1 Which process do plant cells use to capture and store energy from sunlight?

   A endocytosis
   B fermentation
   C glycolysis
   D photosynthesis

Use the diagram below to answer question 2.

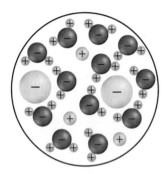
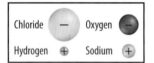

2 The diagram shows salt dissolved in water. What does it show about water molecules and chloride ions?

   A A water molecule consists of oxygen and chloride ions.
   B A water molecule is surrounded by several chloride ions.
   C A water molecule moves away from a chloride ion.
   D A water molecule points its positive end toward a chloride ion.

3 Which transport process requires the use of a cell's energy?

   A diffusion
   B osmosis
   C active transport
   D facilitated diffusion

4 Diffusion differs from active cell transport processes because it

   A forces large molecules from a cell.
   B keeps a cell's boundary intact.
   C moves substances into a cell.
   D needs none of a cell's energy.

Use the diagram below to answer questions 5 and 6.

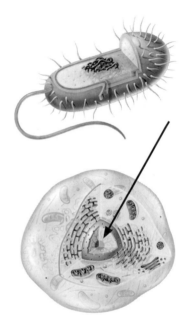

5 Which structure does the arrow point to in the eukaryotic cell?

   A cytoplasm
   B lysosome
   C nucleus
   D ribosome

6 Which feature does a typical prokaryotic cell have that is missing from some eukaryotic cells, like the one above?

   A cytoplasm
   B DNA
   C cell membrane
   D cell wall

Standardized Test Practice

7 Which explains why the ratio of cell surface area to volume affects the cell size? Cells with a high surface-to-volume ratio

   A   consume energy efficiently.

   B   produce waste products slowly.

   C   suffer from diseases frequently.

   D   transport substances effectively.

Use the diagram below to answer question 8.

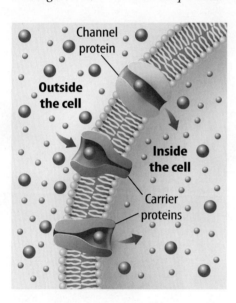

8 Which statement is NOT true of carrier proteins and channel proteins?

   A   Carrier proteins change shape as they function but channel proteins do not.

   B   Carrier proteins and channel proteins extend through the cell membrane.

   C   Channel proteins move items inside a cell but carrier proteins do not.

   D   Channel proteins and carrier proteins perform facilitated diffusion.

Constructed Response

9 Copy the table below and complete it using these terms: *cell membrane, cell wall, chloroplast, cytoplasm, cytoskeleton, nucleus.*

Cell Structure	Function
	Maintains the shape of an animal cell
	Controls the activities of a cell
	Traps energy from the Sun
	Controls the materials going in and out of a cell
	Holds the structures of a cell in a watery mix
	Maintains the shape of some plant cells

10 Name the kinds of organisms that have cells with cell walls. Name the kinds of organisms that have cells without cell walls. Briefly describe the benefits of cell walls for organisms.

11 Draw simple diagrams of an animal cell and a plant cell. Label the nucleus, the cytoplasm, the mitochondria, the cell membrane, the chloroplasts, the cell wall, and the central vacuole in the appropriate cells. Briefly describe the main differences between the two cells.

NEED EXTRA HELP?											
If You Missed Question...	1	2	3	4	5	6	7	8	9	10	11
Go to Lesson...	4	1	3	3	2	2	3	3	2	2	2

Chapter 3
From a Cell to an Organism

 How can one cell become a multicellular organism?

Inquiry What's happening inside?

From the outside, a chicken egg looks like a simple oval object. But big changes are taking place inside the egg. Over several weeks, the one cell in the egg will grow and divide and become a chick.

- How did the original cell change over time?
- What might have happened to the chick's cells as the chick grew?
- How can one cell become a multicellular chick?

Get Ready to Read

What do you think?
Before you read, decide if you agree or disagree with each of these statements. As you read this chapter, see if you change your mind about any of the statements.

1. Cell division produces two identical cells.
2. Cell division is important for growth.
3. At the end of the cell cycle, the original cell no longer exists.
4. Unicellular organisms do not have all the characteristics of life.
5. All the cells in a multicellular organism are the same.
6. Some organs work together as part of an organ system.

ConnectED Your one-stop **online resource**

connectED.mcgraw-hill.com

- Video
- WebQuest
- Audio
- Assessment
- Review
- Concepts in Motion
- Inquiry
- Multilingual eGlossary

Lesson 1

Reading Guide

Key Concepts
ESSENTIAL QUESTIONS

- What are the phases of the cell cycle?
- Why is the result of the cell cycle important?

Vocabulary

cell cycle p. 85
interphase p. 86
sister chromatid p. 88
centromere p. 88
mitosis p. 89
cytokinesis p. 89
daughter cell p. 89

 Multilingual eGlossary

 Video BrainPOP®

The Cell Cycle and Cell Division

Inquiry Time to Split?

Unicellular organisms such as these reproduce when one cell divides into two new cells. The two cells are identical to each other. What do you think happened to the contents of the original cell before it divided?

Inquiry Launch Lab

15 minutes

Why isn't your cell like mine?

All living things are made of cells. Some are made of only one cell, while others are made of trillions of cells. Where do all those cells come from?

1. Read and complete a lab safety form.
2. Ask your team members to face away from you. Draw an animal cell on a sheet of **paper.** Include as many organelles as you can.
3. Use **scissors** to cut the cell drawing into equal halves. Fold each sheet of paper in half so the drawing cannot be seen.
4. Ask your team members to face you. Give each team member half of the cell drawing.
5. Have team members sit facing away from each other. Each person should use a **glue stick** to attach the cell half to one side of a sheet of paper. Then, each person should draw the missing cell half.
6. Compare the two new cells to your original cell.

Think About This

1. How did the new cells compare to the original cell?
2. 🔑 **Key Concept** What are some things that might be done in the early steps to produce two new cells that are more like the original cell?

The Cell Cycle

No matter where you live, you have probably noticed that the weather changes in a regular pattern each year. Some areas experience four seasons—winter, spring, summer, and fall. In other parts of the world, there are only two seasons—rainy and dry. As seasons change, temperature, precipitation, and the number of hours of sunlight vary in a regular cycle.

These changes can affect the life cycles of organisms such as trees. Notice how the tree in **Figure 1** changes with the seasons. Like changing seasons or the growth of trees, cells go through cycles. *Most cells in an organism go through a cycle of growth, development, and division called the* **cell cycle.** Through the cell cycle, organisms grow, develop, replace old or damaged cells, and produce new cells.

Figure 1 This maple tree changes in response to a seasonal cycle.

✓ **Visual Check** List the seasonal changes of this maple tree.

FOLDABLES

Make a folded book from a sheet of paper. Label the front *The Cell Cycle*, and label the inside of the book as shown. Open the book completely and use the full sheet to illustrate the cell cycle.

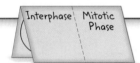

Phases of the Cell Cycle

There are two main phases in the cell cycle—interphase and the mitotic (mi TAH tihk) phase. **Interphase** *is the period during the cell cycle of a cell's growth and development.* A cell spends most of its life in interphase, as shown in **Figure 2**. During interphase, most cells go through three stages:

- rapid growth and replication, or copying, of the membrane-bound structures called organelles;
- copying of DNA, the genetic information in a cell; and
- preparation for cell division.

Interphase is followed by a shorter period of the cell cycle known as the mitotic phase. A cell reproduces during this phase. The mitotic phase has two stages, as illustrated in **Figure 2**. The nucleus divides in the first stage, and the cell's fluid, called the cytoplasm, divides in the second stage. The mitotic phase creates two new identical cells. At the end of this phase, the original cell no longer exists.

 Key Concept Check What are the two main phases of the cell cycle?

The Cell Cycle

Figure 2 A cell spends most of its life growing and developing during interphase.

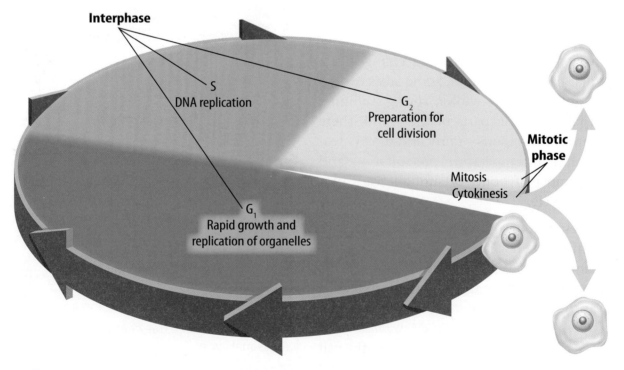

Visual Check Which stage of interphase is the longest?

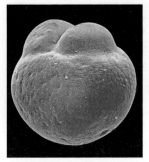
2-cell stage
SEM Magnification: 160×

4-cell stage
SEM Magnification: 155×

32-cell stage
SEM Magnification: 150×

256-cell stage
SEM Magnification: 130×

▲ **Figure 3** The fertilized egg of a zebra fish divides into 256 cells in 2.5 hours.

Length of a Cell Cycle

The time it takes a cell to complete the cell cycle depends on the type of cell that is dividing. Recall that a eukaryotic cell has membrane-bound organelles, including a nucleus. For some eukaryotic cells, the cell cycle might last only eight minutes. For other cells, the cycle might take as long as one year. Most dividing human cells normally complete the cell cycle in about 24 hours. As illustrated in **Figure 3**, the cells of some organisms divide very quickly.

REVIEW VOCABULARY
eukaryotic
a cell with membrane-bound structures

Interphase

As you have read, interphase makes up most of the cell cycle. Newly produced cells begin interphase with a period of rapid growth—the cell gets bigger. This is followed by cellular activities such as making proteins. Next, actively dividing cells make copies of their DNA and prepare for cell division. During interphase, the DNA is called chromatin (KROH muh tun). Chromatin is long, thin strands of DNA, as shown in **Figure 4**. When scientists dye a cell in interphase, the nucleus looks like a plate of spaghetti. This is because the nucleus contains many strands of chromatin tangled together.

Figure 4 During interphase, the nuclei of an animal cell and a plant cell contain long, thin strands of DNA called chromatin. ▼

Interphase

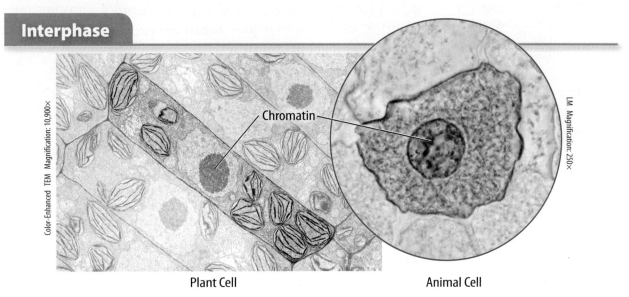

Plant Cell

Animal Cell

Lesson 1
EXPLAIN

(l)Dr. Richard Kessel & Dr. Gene Shih/Visuals Unlimited.

Table 1 Phases of the Cell Cycle — Concepts in Motion — Interactive Table

Phase	Stage	Description
Interphase	G_1	growth and cellular functions; organelle replication
Interphase	S	growth and chromosome replication; organelle replication
Interphase	G_2	growth and cellular functions; organelle replication
Mitotic phase	mitosis	division of nucleus
Mitotic phase	cytokinesis	division of cytoplasm

▲ **Table 1** The two phases of the cell cycle can each be divided into different stages.

Phases of Interphase

Scientists divide interphase into three stages, as shown in **Table 1.** Interphase begins with a period of rapid growth—the G_1 stage. This stage lasts longer than other stages of the cell cycle. During G_1, a cell grows and carries out its normal cell functions. For example, during G_1, cells that line your stomach make enzymes that help digest your food. Although most cells continue the cell cycle, some cells stop the cell cycle at this point. For example, mature nerve cells in your brain remain in G_1 and do not divide again.

During the second stage of interphase—the S stage—a cell continues to grow and copies its DNA. There are now identical strands of DNA. These identical strands of DNA ensure that each new cell gets a copy of the original cell's genetic information. Each strand of DNA coils up and forms a chromosome. Identical chromosomes join together. The cell's DNA is now arranged as pairs of identical chromosomes. Each pair is called a duplicated chromosome. *Two identical chromosomes, called* **sister chromatids,** *make up a duplicated chromosome,* as shown in **Figure 5.** Notice that the *sister chromatids are held together by a structure called the* **centromere.**

The final stage of interphase—the G_2 stage—is another period of growth and the final preparation for the mitotic phase. A cell uses energy copying DNA during the S stage. During G_2, the cell stores energy that will be used during the mitotic phase of the cell cycle.

✓ **Reading Check** Describe what happens in the G_2 phase.

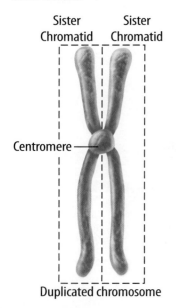

Figure 5 The coiled DNA forms a duplicated chromosome made of two sister chromatids connected at the centromere. ▼

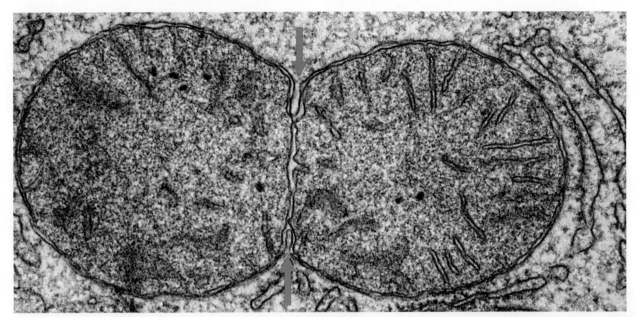

TEM Magnification: Unavailable

Organelle Replication

During cell division, the organelles in a cell are distributed between the two new cells. Before a cell divides, it makes a copy of each organelle. This enables the two new cells to function properly. Some organelles, such as the energy-processing mitochondria and chloroplasts, have their own DNA. These organelles can make copies of themselves on their own, as shown in **Figure 6.** A cell produces other organelles from materials such as proteins and lipids. A cell makes these materials using the information contained in the DNA inside the nucleus. Organelles are copied during all stages of interphase.

The Mitotic Phase

The mitotic phase of the cell cycle follows interphase. It consists of two stages: mitosis (mi TOH sus) and cytokinesis (si toh kuh NEE sus). *In* **mitosis,** *the nucleus and its contents divide. In* **cytokinesis,** *the cytoplasm and its contents divide.* **Daughter cells** *are the two new cells that result from mitosis and cytokinesis.*

During mitosis, the contents of the nucleus divide, forming two identical nuclei. The sister chromatids of the duplicated chromosomes separate from each other. This gives each daughter cell the same genetic information. For example, a cell that has ten duplicated chromosomes actually has 20 chromatids. When the cell divides, each daughter cell will have ten different chromatids. Chromatids are now called chromosomes.

In cytokinesis, the cytoplasm divides and forms the two new daughter cells. Organelles that were made during interphase are divided between the daughter cells.

Figure 6 This mitochondrion is in the final stage of dividing.

Word Origin

mitosis
from Greek *mitos*, means "warp thread"; and Latin *-osis*, means "process"

Phases of Mitosis

Like interphase, mitosis is a continuous process that scientists divide into different phases, as shown in **Figure 7**.

Prophase During the first phase of mitosis, called prophase, the copied chromatin coils together tightly. The coils form visible duplicated chromosomes. The nucleolus disappears, and the nuclear membrane breaks down. Structures called spindle fibers form in the cytoplasm.

Metaphase During metaphase, the spindle fibers pull and push the duplicated chromosomes to the middle of the cell. Notice in **Figure 7** that the chromosomes line up along the middle of the cell. This arrangement ensures that each new cell will receive one copy of each chromosome. Metaphase is the shortest phase in mitosis, but it must be completed successfully for the new cells to be identical.

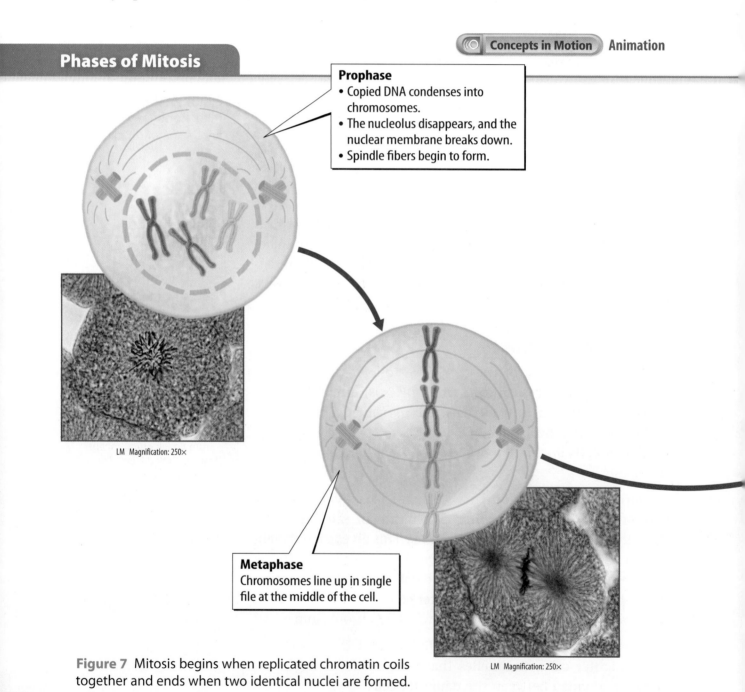

Phases of Mitosis

Prophase
- Copied DNA condenses into chromosomes.
- The nucleolus disappears, and the nuclear membrane breaks down.
- Spindle fibers begin to form.

LM Magnification: 250×

Metaphase
Chromosomes line up in single file at the middle of the cell.

LM Magnification: 250×

Figure 7 Mitosis begins when replicated chromatin coils together and ends when two identical nuclei are formed.

Anaphase In anaphase, the third stage of mitosis, the two sister chromatids in each chromosome separate from each other. The spindle fibers pull them in opposite directions. Once separated, the chromatids are now two identical single-stranded chromosomes. As they move to opposite sides of a cell, the cell begins to get longer. Anaphase is complete when the two identical sets of chromosomes are at opposite ends of a cell.

Telophase During telophase, the spindle fibers begin to disappear. Also, the chromosomes begin to uncoil. A nuclear membrane forms around each set of chromosomes at either end of the cell. This forms two new identical nuclei. Telophase is the final stage of mitosis. It is often described as the reverse of prophase because many of the processes that occur during prophase are reversed during telophase.

 Reading Check What are the phases of mitosis?

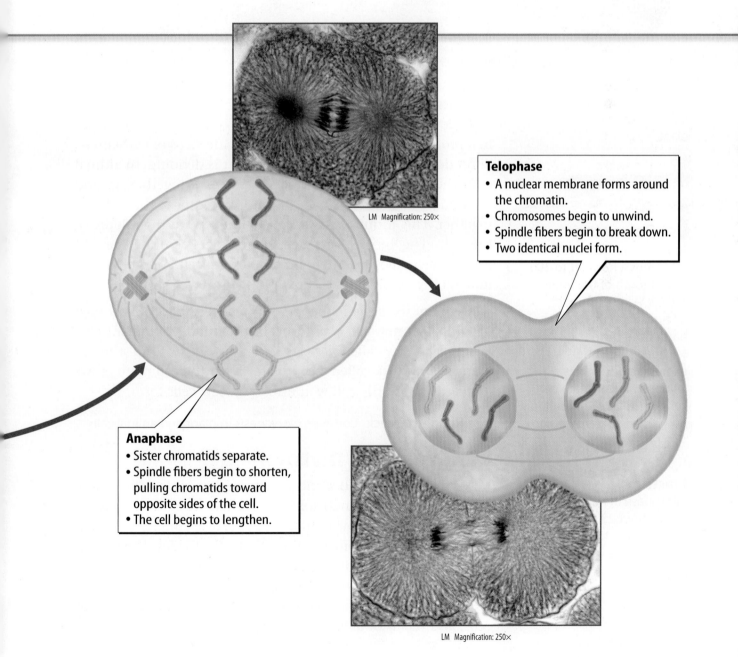

LM Magnification: 250×

Telophase
- A nuclear membrane forms around the chromatin.
- Chromosomes begin to unwind.
- Spindle fibers begin to break down.
- Two identical nuclei form.

Anaphase
- Sister chromatids separate.
- Spindle fibers begin to shorten, pulling chromatids toward opposite sides of the cell.
- The cell begins to lengthen.

LM Magnification: 250×

Cytokinesis

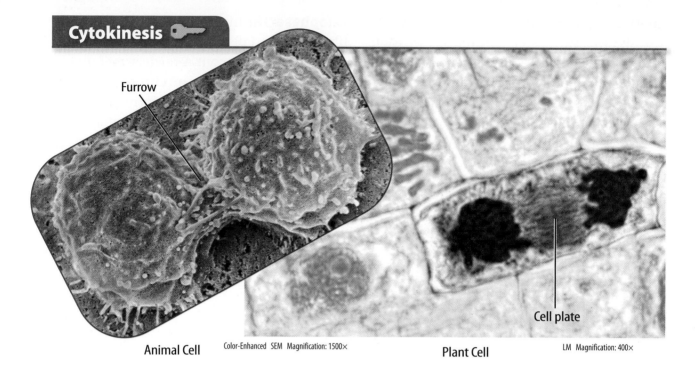

Furrow

Animal Cell Color-Enhanced SEM Magnification: 1500×

Cell plate

Plant Cell LM Magnification: 400×

Figure 8 Cytokinesis differs in animal cells and plant cells.

Dividing the Cell's Components

Following the last phase of mitosis, a cell's cytoplasm divides in a process called cytokinesis. The specific steps of cytokinesis differ depending on the type of cell that is dividing. In animal cells, the cell membrane contracts, or squeezes together, around the middle of the cell. Fibers around the center of the cell pull together. This forms a crease, called a furrow, in the middle of the cell. The furrow gets deeper and deeper until the cell membrane comes together and divides the cell. An animal cell undergoing cytokinesis is shown in **Figure 8.**

Cytokinesis in plants happens in a different way. As shown in **Figure 8,** a new cell wall forms in the middle of a plant cell. First, organelles called vesicles join together to form a membrane-bound disk called a cell plate. Then the cell plate grows outward toward the cell wall until two new cells form.

 Reading Check Compare cytokinesis in plant and animal cells.

Results of Cell Division

Recall that the cell cycle results in two new cells. These daughter cells are genetically identical to each other and to the original cell that no longer exists. For example, a human cell has 46 chromosomes. When that cell divides, it will produce two new cells with 46 chromosomes each. The cell cycle is important for reproduction in some organisms, growth in multicellular organisms, replacement of worn out or damaged cells, and repair of damaged tissues.

Math Skills

Use Percentages
A percentage is a ratio that compares a number to 100. If the length of the entire cell cycle is 24 hours, 24 hours equals 100%. If part of the cycle takes 6.0 hours, it can be expressed as 6.0 hours/24 hours. To calculate percentage, divide and multiply by 100. Add a percent sign.

$$\frac{6.0}{24} = 0.25 \times 100 = 25\%$$

Practice
Interphase in human cells takes about 23 hours. If the cell cycle is 24 hours, what percentage is interphase?

- Math Practice
- Personal Tutor

Reproduction

In some unicellular organisms, cell division is a form of reproduction. For example, an organism called a paramecium often reproduces by dividing into two new daughter cells or two new paramecia. Cell division is also important in other methods of reproduction in which the offspring are identical to the parent organism.

Growth

Cell division allows multicellular organisms, such as humans, to grow and develop from one cell (a fertilized egg). In humans, cell division begins about 24 hours after fertilization and continues rapidly during the first few years of life. It is likely that during the next few years you will go through another period of rapid growth and development. This happens because cells divide and increase in number as you grow and develop.

Replacement

Even after an organism is fully grown, cell division continues. It replaces cells that wear out or are damaged. The outermost layer of your skin is always rubbing or flaking off. A layer of cells below the skin's surface is constantly dividing. This produces millions of new cells daily to replace the ones that are rubbed off.

Repair

Cell division is also critical for repairing damage. When a bone breaks, cell division produces new bone cells that patch the broken pieces back together.

Not all damage can be repaired, however, because not all cells continue to divide. Recall that mature nerve cells stop the cell cycle in interphase. For this reason, injuries to nerve cells often cause permanent damage.

 Key Concept Check Why is the result of the cell cycle important?

Inquiry MiniLab 20 minutes

How does mitosis work?

The dolix is a mythical animal whose cells contain just two chromosomes. What happens to a dolix cell nucleus during mitosis?

1. Read and complete a lab safety form.
2. Form four 60-cm lengths of **yarn** into large circles on four separate sheets of **paper.** Each piece of paper represents one phase of mitosis, and the yarn represents the cell membrane.
3. On each sheet of paper, model one phase of mitosis using different colors of yarn to represent the nuclear membrane, the spindles, and the chromosomes. Use **twist ties** to represent centromeres. **Tape** the yarn in place.
4. Label your models, or develop a key to indicate which color is used for which part.

Analyze and Conclude

1. **Identify** If you were to model a dolix cell's nucleus before mitosis began, what would your model look like? Would you be able to see the individual chromosomes?
2. **Integrate** What would a model of your cell look like during the stage immediately following mitosis? What is this stage?
3. **Key Concept** During mitosis, a cell forms two new, identical nuclei. Use your models to explain why, in order to do this, mitosis must occur after events in interphase.

Lesson 1 Review

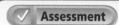 Assessment Online Quiz

Visual Summary

During interphase, most cells go through periods of rapid growth and replication of organelles, copying DNA, and preparation for cell division.

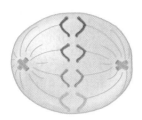

The nucleus and its contents divide during mitosis.

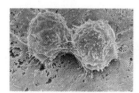

The cytoplasm and its contents divide during cytokinesis.

FOLDABLES

Use your lesson Foldable to review the lesson. Save your Foldable for the project at the end of the chapter.

What do you think NOW?

You first read the statements below at the beginning of the chapter.

1. Cell division produces two identical cells.
2. Cell division is important for growth.
3. At the end of the cell cycle, the original cell no longer exists.

Did you change your mind about whether you agree or disagree with the statements? Rewrite any false statements to make them true.

Use Vocabulary

1. **Distinguish** between mitosis and cytokinesis.
2. A duplicated chromosome is made of two _____.
3. **Use the term** *interphase* in a sentence.

Understand Key Concepts

4. Which is NOT part of mitosis?
   A. anaphase   C. prophase
   B. interphase   D. telophase
5. **Construct** a table to show the different phases of mitosis and what happens during each.
6. **Give three examples** of why the result of the cell cycle is important.

Interpret Graphics

7. **Identify** The animal cell on the right is in what phase of mitosis? Explain your answer.

8. **Organize** Copy and fill in the graphic organizer below to show the results of cell division.

Critical Thinking

9. **Predict** what might happen to a cell if it were unable to divide by mitosis.

Math Skills

— Math Practice —

10. The mitotic phase of the human cell cycle takes approximately 1 hour. What percentage of the 24-hour cell cycle is the mitotic phase?

DNA Fingerprinting

SCIENCE & SOCIETY

▼ DNA

Solving Crimes One Strand at a Time

Every cell in your body has the same DNA in its nucleus. Unless you are an identical twin, your DNA is entirely unique. Identical twins have identical DNA because they begin as one cell that divides and separates. When your cells begin mitosis, they copy their DNA. Every new cell has the same DNA as the original cells. That is why DNA can be used to identify people. Just as no two people have the same fingerprints, your DNA belongs to you alone.

Using scientific methods to solve crimes is called forensics. DNA fingerprinting is now a basic tool in forensics. Samples collected from a crime scene can be compared to millions of samples previously collected and indexed in a computer.

Every day, everywhere you go, you leave a trail of DNA. It might be in skin cells. It might be in hair or in the saliva you used to lick an envelope. If you commit a crime, you will most likely leave DNA behind. An expert crime scene investigator will know how to collect that DNA.

DNA evidence can prove innocence as well. Investigators have reexamined DNA found at old crime scenes. Imprisoned persons have been proven not guilty through DNA fingerprinting methods that were not yet available when a crime was committed.

DNA fingerprinting can also be used to identify bodies that had previously been known only as a John or Jane Doe.

▼ The Federal Bureau of Investigation (FBI) has a nationwide index of DNA samples called CODIS (Combined DNA Index System).

It's Your Turn

DISCOVER Your cells contain organelles called mitochondria. They have their own DNA, called mitochondrial DNA. Your mitochondrial DNA is identical to your mother's mitochondrial DNA. Find out how this information is used.

Lesson 1 EXTEND

Lesson 2

Levels of Organization

Reading Guide

Key Concepts
ESSENTIAL QUESTIONS

- How do unicellular and multicellular organisms differ?
- How does cell differentiation lead to the organization within a multicellular organism?

Vocabulary

cell differentiation p. 99
stem cell p. 100
tissue p. 101
organ p. 102
organ system p. 103

Multilingual eGlossary

Video BrainPOP®

Inquiry Scales on Wings?

This butterfly has a distinctive pattern of colors on its wings. The pattern is formed by clusters of tiny scales. In a similar way, multicellular organisms are made of many small parts working together.

Inquiry Launch Lab

15 minutes

How is a system organized?

The places people live are organized in a system. Do you live in or near a city? Cities contain things such as schools and stores that enable them to function on their own. Many cities together make up another level of organization.

1. Read and complete a lab safety form.
2. Using a **metric ruler** and **scissors,** measure and cut squares of **construction paper** that are 4 cm, 8 cm, 12 cm, 16 cm, and 20 cm on each side. Use a different color for each square.
3. Stack the squares from largest to smallest, and glue them together.
4. Cut apart the *City, Continent, Country, County,* and *State* labels your teacher gives you.
5. Use a **glue stick** to attach the *City* label to the smallest square. Sort the remaining labels from smallest to largest, and glue to the corresponding square.

Think About This

1. What is the largest level of organization a city belongs to?
2. Can any part of the system function without the others? Explain.
3. **Key Concept** How do you think the system used to organize where people live is similar to how your body is organized?

Life's Organization

You might recall that all matter is made of atoms and that atoms combine and form molecules. Molecules make up cells. A large animal, such as a Komodo dragon, is not made of one cell. Instead, it is composed of trillions of cells working together. Its skin, shown in **Figure 9,** is made of many cells that are specialized for protection. The Komodo dragon has other types of cells, such as blood cells and nerve cells, that perform other functions. Cells work together in the Komodo dragon and enable it to function. In the same way, cells work together in you and in other multicellular organisms.

Recall that some organisms are made of only one cell. These unicellular organisms carry out all the activities necessary to survive, such as absorbing nutrients and getting rid of wastes. But no matter their sizes, all organisms are made of cells.

Color-Enhanced SEM Magnification: 12×

Figure 9 Skin cells are only one of the many kinds of cells that make up a Komodo dragon.

Unicellular Organisms

Figure 10 Unicellular organisms carry out life processes within one cell.

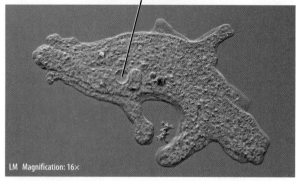

This unicellular amoeba captures a desmid for food.

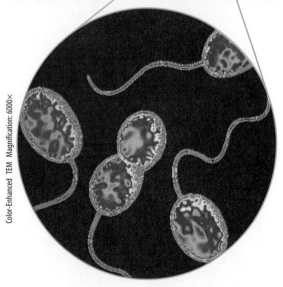

These heat-loving bacteria are often found in hot springs as shown here. They get their energy to produce food from sulfur instead of from light like plants.

Unicellular Organisms

As you read on the previous page, some organisms have only one cell. Unicellular organisms do all the things needed for their survival within that one cell. For example, the amoeba in **Figure 10** is ingesting another unicellular organism, a type of green algae called a desmid, for food. Unicellular organisms also respond to their environment, get rid of waste, grow, and even reproduce on their own. Unicellular organisms include both prokaryotes and some eukaryotes.

Prokaryotes

Recall that a cell without a membrane-bound nucleus is a prokaryotic cell. In general, prokaryotic cells are smaller than eukaryotic cells and have fewer cell structures. A unicellular organism made of one prokaryotic cell is called a prokaryote. Some prokaryotes live in groups called colonies. Some can also live in extreme environments, as shown in **Figure 10**.

Eukaryotes

You might recall that a eukaryotic cell has a nucleus surrounded by a membrane and many other specialized organelles. For example, the amoeba shown in **Figure 10** has an organelle called a contractile vacuole. It functions like a bucket that is used to bail water out of a boat. A contractile vacuole collects excess water from the amoeba's cytoplasm. Then it pumps the water out of the amoeba. This prevents the amoeba from swelling and bursting.

A unicellular organism that is made of one eukaryotic cell is called a eukaryote. There are thousands of different unicellular eukaryotes, such as algae that grow on the inside of an aquarium and the fungus that causes athlete's foot.

Reading Check Give an example of a unicellular eukaryotic organism.

Multicellular Organisms

Multicellular organisms are made of many eukaryotic cells working together, like the crew on an airplane. Each member of the crew, from the pilot to the mechanic, has a specific job that is important for the plane's operation. Similarly, each type of cell in a multicellular organism has a specific job that is important to the survival of the organism.

 Key Concept Check How do unicellular and multicellular organisms differ?

Cell Differentiation

As you read in the last lesson, all cells in a multicellular organism come from one cell—a fertilized egg. Cell division starts quickly after fertilization. The first cells made can become any type of cell, such as a muscle cell, a nerve cell, or a blood cell. *The process by which cells become different types of cells is called* **cell differentiation** (dihf uh ren shee AY shun).

You might recall that a cell's instructions are contained in its chromosomes. Also, nearly all the cells of an organism have identical sets of chromosomes. If an organism's cells have identical sets of instructions, how can cells be different? Different cell types use different parts of the instructions on the chromosomes. A few of the many different types of cells that can result from human cell differentiation are shown in **Figure 11.**

FOLDABLES
Make a layered book from three sheets of notebook paper. Label it as shown. Use your book to describe the levels of organization that make up organisms.

Levels of Organization
Cell
Tissue
Organ
Organ System
Organism

Figure 11 A fertilized egg produces cells that can differentiate into a variety of cell types.

Review   Personal Tutor

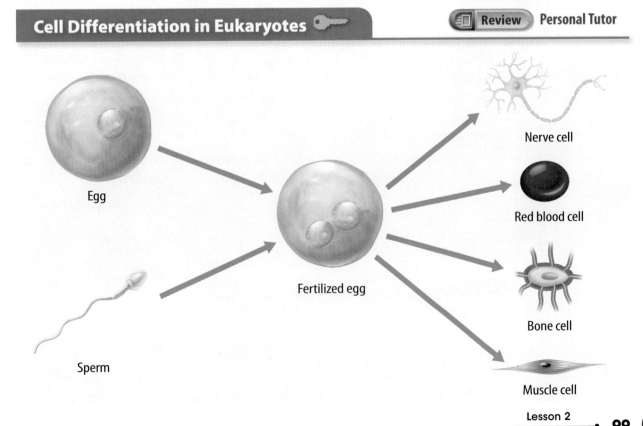

Cell Differentiation in Eukaryotes

Animal Stem Cells Not all cells in a developing animal differentiate. **Stem cells** *are unspecialized cells that are able to develop into many different cell types.* There are many stem cells in embryos but fewer in adult organisms. Adult stem cells are important for the cell repair and replacement you read about in Lesson 1. For example, stem cells in your bone marrow can produce more than a dozen different types of blood cells. These replace ones that are damaged or worn out. Stem cells have also been discovered in skeletal muscles. These stem cells can produce new muscle cells when the fibers that make up the muscle are torn.

Plant Cells Plants also have unspecialized cells similar to animal stem cells. These cells are grouped in areas of a plant called meristems (MER uh stemz). Meristems are in different areas of a plant, including the tips of roots and stems, as shown in **Figure 12**. Cell division in meristems produces different types of plant cells with specialized structures and functions, such as transporting materials, making food, storing food, or protecting the plant. These cells might become parts of stems, leaves, flowers, or roots.

SCIENCE USE V. COMMON USE

fiber
Science Use a long muscle cell

Common Use a thread

Figure 12 Plant meristems produce cells that can become part of stems, leaves, flowers, or roots.

Tissues

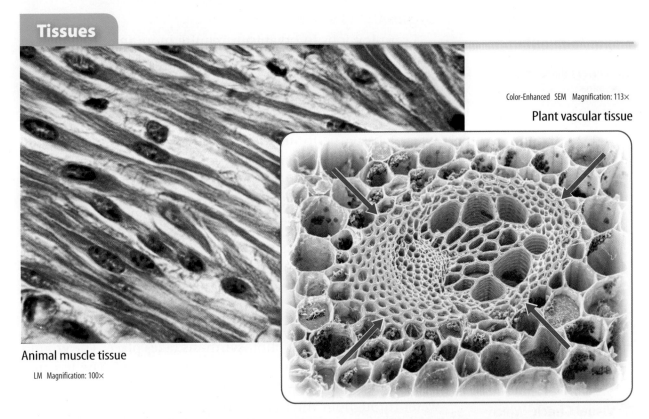

Color-Enhanced SEM Magnification: 113×
Plant vascular tissue

Animal muscle tissue
LM Magnification: 100×

Figure 13 Similar cells work together and form tissues such as this animal muscle tissue that contracts the stomach to help digestion. Plant vascular tissue, indicated by red arrows, moves water and nutrients throughout a plant.

Tissues

In multicellular organisms, similar types of cells are organized into groups. **Tissues** *are groups of similar types of cells that work together to carry out specific tasks.* Humans, like most other animals, have four main types of tissue—muscle, connective, nervous, and epithelial (eh puh THEE lee ul). For example, the animal tissue shown in **Figure 13** is smooth muscle tissue that is part of the stomach. Muscle tissue causes movement. Connective tissue provides structure and support and often connects other types of tissue together. Nervous tissue carries messages to and from the brain. Epithelial tissue forms the protective outer layer of the skin and the lining of major organs and internal body cavities.

Plants also have different types of tissues. The three main types of plant tissue are dermal, vascular (VAS kyuh lur), and ground tissue. Dermal tissue provides protection and helps reduce water loss. Vascular tissue, shown in **Figure 13**, transports water and nutrients from one part of a plant to another. Ground tissue provides storage and support and is where photosynthesis takes place.

Reading Check Compare animal and plant tissues.

> **WORD ORIGIN**
> tissue
> from Latin *texere*, means "weave"

Lesson 2
EXPLAIN

ACADEMIC VOCABULARY
complex
(adjective) made of two or more parts

Organs

Complex jobs in organisms require more than one type of tissue. **Organs** *are groups of different tissues working together to perform a particular job.* For example, your stomach is an organ specialized for breaking down food. It is made of all four types of tissue: muscle, epithelial, nervous, and connective. Each type of tissue performs a specific function necessary for the stomach to work properly. Layers of muscle tissue contract and break up pieces of food, epithelial tissue lines the stomach, nervous tissue sends signals to indicate the stomach is full, and connective tissue supports the stomach wall.

Plants also have organs. The leaves shown in **Figure 14** are organs specialized for photosynthesis. Each leaf is made of dermal, ground, and vascular tissues. Dermal tissue covers the outer surface of a leaf. The leaf is a vital organ because it contains ground tissue that produces food for the rest of the plant. Ground tissue is where photosynthesis takes place. The ground tissue is tightly packed on the top half of a leaf. The vascular tissue moves both the food produced by photosynthesis and water throughout the leaf and the rest of the plant.

Reading Check List the tissues in a leaf organ.

Figure 14 A plant leaf is an organ made of several different tissues.

 Visual Check Which plant tissue makes up the thinnest layer?

Organ Systems

Usually organs do not function alone. Instead, **organ systems** *are groups of different organs that work together to complete a series of tasks*. Human organ systems can be made of many different organs working together. For example, the human digestive system is made of many organs, including the stomach, the small intestine, the liver, and the large intestine. These organs and others all work together to break down food and take it into the body. Blood absorbs and transports nutrients from broken down food to cells throughout the body.

Plants have two major organ systems—the shoot system and the root system. The shoot system includes leaves, stems, and flowers. Food and water are transported throughout the plant by the shoot system. The root system anchors the plant and takes in water and nutrients.

 Reading Check What are the major organ systems in plants?

Inquiry MiniLab
25 minutes

How do cells work together to make an organism?

In a multicellular organism, similar cells work together and make a tissue. A tissue can perform functions that individual cells cannot. Tissues are organized into organs, then organ systems, then organisms. How can you model the levels of organization in an organism?

1. Read and complete a lab safety form.
2. Your teacher will give you a **cardboard shape, macaroni,** and a **permanent marker.**
3. The macaroni represent cells. Use the marker to draw a small circle on each piece of macaroni. This represents the nucleus.
4. Arrange and **glue** enough macaroni on the blank side of the cardboard shape to cover it. Your group of similar cells represents a tissue.
5. One of the squares on the back of your shape is labeled *A, B, C,* or *D*. Find the group with a matching letter. Line up these squares, and use **tape** to connect the two tissues. This represents an organ.
6. Repeat step 4 with the squares labeled *E* or *F*. This represents an organ system.
7. Connect the organ systems by aligning the squares labeled *G* to represent an organism.

Analyze and Conclude

1. Each group had to work with other groups to make a model of an organism. Do cells, tissues, and organs need to work together in organisms? Explain.
2. **Key Concept** How does your model show the levels of organization in living things?

Organisms

Multicellular organisms usually have many organ systems. These systems work together to carry out all the jobs needed for the survival of the organisms. For example, the cells in the leaves and the stems of a plant need water to live. They cannot absorb water directly. Water diffuses into the roots and is transported through the stem to the leaves by the transport system.

In the human body, there are many major organ systems. Each organ system depends on the others and cannot work alone. For example, the cells in the muscle tissue of the stomach cannot survive without oxygen. The stomach cannot get oxygen without working together with the respiratory and circulatory systems. **Figure 15** will help you review how organisms are organized.

Key Concept Check How does cell differentiation lead to the organization within a multicellular organism?

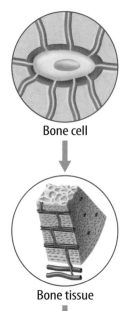

Bone cell

Bone tissue

Bone (organ)

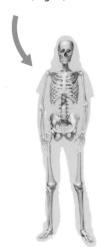

Skeletal system

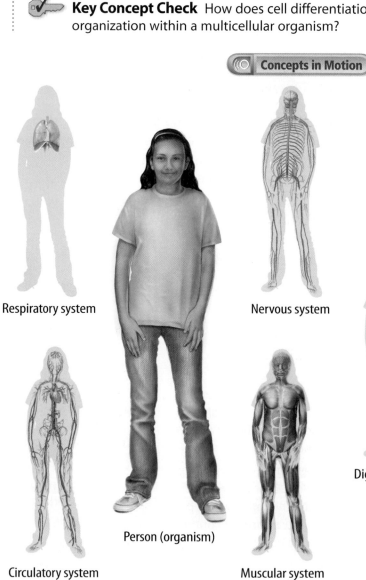

Respiratory system

Nervous system

Digestive system

Circulatory system

Person (organism)

Muscular system

Figure 15 An organism is made of organ systems, organs, tissues, and cells that all function together and enable the organism's survival.

Lesson 2 Review

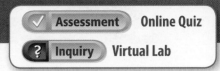

Visual Summary

A unicellular organism carries out all the activities necessary for survival within one cell.

Cells become specialized in structure and function during cell differentiation.

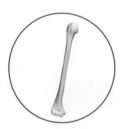

Organs are groups of different tissues that work together to perform a job.

FOLDABLES

Use your lesson Foldable to review the lesson. Save your Foldable for the project at the end of the chapter.

What do you think NOW?

You first read the statements below at the beginning of the chapter.

4. Unicellular organisms do not have all the characteristics of life.
5. All the cells in a multicellular organism are the same.
6. Some organs work together as part of an organ system.

Did you change your mind about whether you agree or disagree with the statements? Rewrite any false statements to make them true.

Use Vocabulary

1. **Define** *cell differentiation* in your own words.
2. **Distinguish** between an organ and an organ system.

Understand Key Concepts

3. **Explain** the difference between a unicellular organism and a multicellular organism.
4. **Describe** how cell differentiation produces different types of cells in animals.
5. Which is the correct sequence of the levels of organization?
   A. cell, organ, tissue, organ system, organism
   B. organism, organ, organ system, tissue, cell
   C. cell, tissue, organ, organ system, organism
   D. tissue, organ, organism, organ system, cell

Interpret Graphics

6. **Organize** Copy and fill in the table below to summarize the characteristics of unicellular and multicellular organisms.

Organism Characteristics	
Unicellular	Multicellular

Critical Thinking

7. **Predict** A mistake occurs during mitosis of a muscle stem cell. How might this affect muscle tissue?
8. **Compare** the functions of a cell to the functions of an organism, such as getting rid of wastes.

Inquiry Lab

90 minutes

Cell Differentiation

Materials

cooked eggs

boiled chicken leg

forceps

dissecting scissors

plastic knife

paper towels

Safety

It's pretty amazing that a whole chicken with wings, feet, beak, feathers, and internal organs can come from one cell, a fertilized egg. Shortly after fertilization, the cell begins to divide. The new cells in the developing embryo become specialized both in structure and function. The process by which cells become specialized is called cellular differentiation.

Question
How does a single cell become a multicellular organism?

Procedure

1. Read and complete a lab safety form.
2. Carefully examine the outside of your egg. Remove the shell.
3. Dissect the egg on a paper towel, cutting it in half from tip to rounded end. Examine the inside.
4. Record your observations in your Science Journal. Include a labeled drawing. Infer the function of each part.
5. Discard all your trash in the container provided.
6. Examine the outside of the chicken leg. Describe the skin and its functions.

7. Carefully remove the skin using forceps and dissecting scissors. Put the skin in your discard container. Now you should see evidence of fat and muscles. You may also be able to see some blood vessels and tendons, but these are not always visible after cooking. Describe each part that you see and explain its function.

8. Peel back the muscles to reveal the bones. Tendons, ligaments, and cartilage holding the bones in place may also be evident.

9. Put all your trash in the discard container. Your teacher will give you instructions about cleaning up.

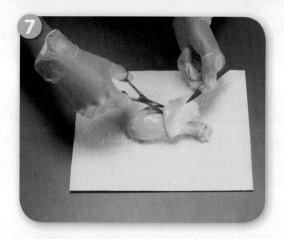

Analyze and Conclude

10. **The Big Idea** A single cell can become a multicellular organism through the process of cell differentiation. How do the organization of the egg and the chicken leg compare?

11. **Summarize** How many different types of cell differentiation did you observe in the chicken leg?

Lab Tips

☑ Work slowly and carefully on your dissections so as not to destroy any structures. Report any accidents to your teacher immediately. Cleaning up is important!

Communicate Your Results

Make a poster about how an egg transforms into a chicken through the process of cell differentiation.

Inquiry Extension

Examine a whole raw chicken or a raw chicken leg that is still attached to a thigh. You might be able to move the muscles in the legs or wings and see parts that were not visible in this lab. Be sure to wear gloves and to wash well with soap and water after touching the raw chicken.

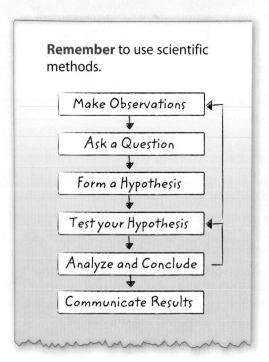

Remember to use scientific methods.

- Make Observations
- Ask a Question
- Form a Hypothesis
- Test your Hypothesis
- Analyze and Conclude
- Communicate Results

Chapter 3 Study Guide

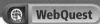

THE BIG IDEA: Through cell division, one cell can produce new cells to grow and develop into a multicellular organism.

Key Concepts Summary

Lesson 1: The Cell Cycle and Cell Division

- The **cell cycle** consists of two phases. During **interphase,** a cell grows and its chromosomes and organelles replicate. During the mitotic phase of the cell cycle, the nucleus divides during **mitosis,** and the cytoplasm divides during **cytokinesis.**
- The cell cycle results in two genetically identical **daughter cells.** The original parent cell no longer exists.
- The cell cycle is important for growth in multicellular organisms, reproduction in some organisms, replacement of worn-out cells, and repair of damaged cells.

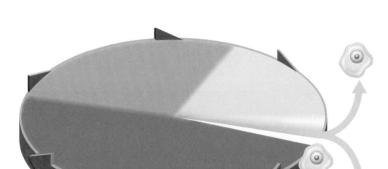

Vocabulary

cell cycle p. 85
interphase p. 86
sister chromatid p. 88
centromere p. 88
mitosis p. 89
cytokinesis p. 89
daughter cell p. 89

Lesson 2: Levels of Organization

- The one cell of a unicellular organism is able to obtain all the materials that it needs to survive.
- In a multicellular organism, cells cannot survive alone and must work together to provide the organism's needs.
- Through **cell differentiation,** cells become different types of cells with specific functions. Cell differentiation leads to the formation of **tissues, organs,** and **organ systems.**

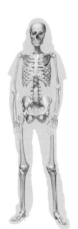

cell differentiation p. 99
stem cell p. 100
tissue p. 101
organ p. 102
organ system p. 103

Study Guide

Review
- Personal Tutor
- Vocabulary eGames
- Vocabulary eFlashcards

FOLDABLES Chapter Project

Assemble your lesson Foldables as shown to make a Chapter Project. Use the project to review what you have learned in this chapter.

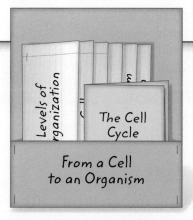

Use Vocabulary

1. Use the term *sister chromatids* in a sentence.

2. Define the term *centromere* in your own words.

3. The new cells formed by mitosis are called _____.

4. Use the term *cell differentiation* in a sentence.

5. Define the term *stem cell* in your own words.

6. Organs are groups of _____ working together to perform a specific task.

Link Vocabulary and Key Concepts

Concepts in Motion — Interactive Concept Map

Copy this concept map, and then use vocabulary terms from the previous page and from the chapter to complete the concept map.

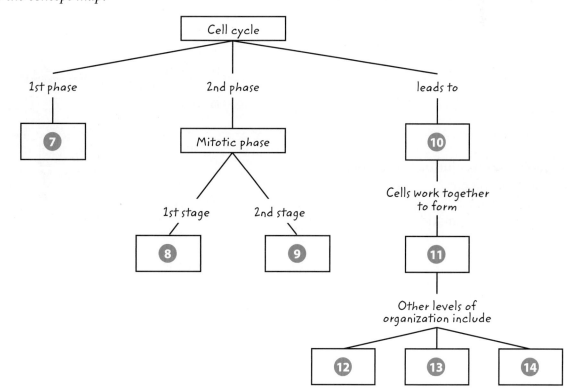

Chapter 3 Study Guide • 109

Chapter 3 Review

Understand Key Concepts

1. Chromosomes line up in the center of the cell during which phase?
   A. anaphase
   B. metaphase
   C. prophase
   D. telophase

2. Which stage of the cell cycle precedes cytokinesis?
   A. G_1
   B. G_2
   C. interphase
   D. mitosis

Use the figure below to answer questions 3 and 4.

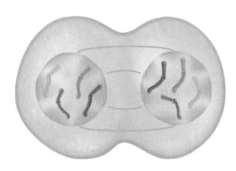

3. The figure represents which stage of mitosis?
   A. anaphase
   B. metaphase
   C. prophase
   D. telophase

4. What forms during this phase?
   A. centromere
   B. furrow
   C. sister chromatid
   D. two nuclei

5. What is the longest part of the cell cycle?
   A. anaphase
   B. cytokinesis
   C. interphase
   D. mitosis

6. A plant's root system is which level of organization?
   A. cell
   B. organ
   C. organ system
   D. tissue

7. Where is a meristem often found?
   A. liver cells
   B. muscle tissue
   C. tip of plant root
   D. unicellular organism

8. Which is NOT a type of human tissue?
   A. connective
   B. meristem
   C. muscle
   D. nervous

9. Which are unspecialized cells?
   A. blood cells
   B. muscle cells
   C. nerve cells
   D. stem cells

10. Which level of organization is shown in the figure below?
    A. cell
    B. organ
    C. organ system
    D. tissue

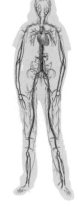

11. Which level of organization completes a series of tasks?
    A. cell
    B. organ
    C. organ system
    D. tissue

Chapter Review

Assessment
Online Test Practice

Critical Thinking

12 Sequence the events that occur during the phases of mitosis.

13 Infer why the chromatin condenses into chromosomes before mitosis begins.

14 Create Use the figure below to create a cartoon that shows a duplicated chromosome separating into two sister chromatids.

15 Classify a leaf as a tissue or an organ. Explain your choice.

16 Distinguish between a tissue and an organ.

17 Construct a table that lists and defines the different levels of organization.

18 Summarize the differences between unicellular organisms and multicellular organisms.

Writing in Science

19 Write a five-sentence paragraph describing a human organ system. Include a main idea, supporting details, and a concluding statement.

REVIEW THE BIG IDEA

20 Why is cell division important for multicellular organisms?

21 The photo below shows a chick growing inside an egg. An egg begins as one cell. How can one cell become a chick?

Math Skills

Review Math Practice

Use Percentages

22 During an interphase lasting 23 hours, the S stage takes an average of 8.0 hours. What percentage of interphase is taken up by the S stage?

Use the following information to answer questions 23 through 25.

During a 23-hour interphase, the G_1 stage takes 11 hours and the S stage takes 8.0 hours.

23 What percentage of interphase is taken up by the G_1 and S stages?

24 What percentage of interphase is taken up by the G_2 phase?

25 How many hours does the G_2 phase last?

Standardized Test Practice

Record your answers on the answer sheet provided by your teacher or on a sheet of paper.

Multiple Choice

1. Which tissue carries messages to and from the brain?
   A connective
   B epithelial
   C muscle
   D nervous

Use the diagram below to answer question 2.

2. What is indicated by the arrow?
   A centromere
   B chromatid
   C chromosome
   D nucleus

3. In which stage of mitosis do spindle fibers form?
   A anaphase
   B metaphase
   C prophase
   D telophase

4. What structures separate during anaphase?
   A centromeres
   B chromatids
   C nuclei
   D organelles

Use the diagram below to answer question 5.

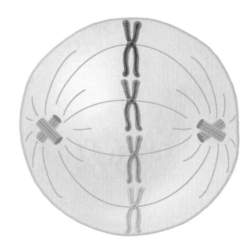

5. What stage of mitosis does the image above represent?
   A anaphase
   B metaphase
   C prophase
   D telophase

6. A plant's dermal tissue
   A produces food for the rest of the plant.
   B provides protection and helps reduce water loss.
   C takes in water and nutrients for use throughout the plant.
   D transports water and nutrients throughout the plant.

7. Which is the most accurate description of a leaf or your stomach?
   A a cell
   B an organ
   C an organ system
   D a tissue

Standardized Test Practice

Use the figure below to answer question 8.

8 Which does this figure illustrate?
  A an organ
  B an organism
  C an organ system
  D a tissue

9 If a cell has 30 chromosomes at the start of mitosis, how many chromosomes will be in each new daughter cell?
  A 10
  B 15
  C 30
  D 60

10 What areas of plants have unspecialized cells?
  A flowers
  B fruits
  C leaves
  D meristems

Constructed Response

Use the figure below to answer questions 11 and 12.

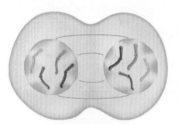

Figure A

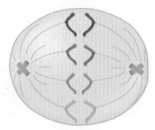

Figure B

11 The figures illustrate two phases of mitosis. Which occurs first: A or B? Explain your reasoning.

12 What stage of the mitotic phase follows those illustrated above? Explain how this stage differs between plant and animal cells.

13 What are some similarities and differences between the G_1 and S stages of interphase?

14 Are all human cells capable of mitosis and cell division? How does this affect the body's ability to repair itself? Support your answer with specific examples.

NEED EXTRA HELP?														
If You Missed Question...	1	2	3	4	5	6	7	8	9	10	11	12	13	14
Go to Lesson...	2	1	1	1	1	2	2	2	1	2	1	1	1	1

Chapter 4
Reproduction of Organisms

 Why do living things reproduce?

Inquiry Time to bond?

Have you ever seen a family of animals, such as the one of penguins shown here? Notice the baby penguin beside its parents. Like all living things, penguins reproduce.

- Do you think all living things have two parents?
- What might happen if the penguins did not reproduce?
- Why do living things reproduce?

Get Ready to Read

What do you think?
Before you read, decide if you agree or disagree with each of these statements. As you read this chapter, see if you change your mind about any of the statements.

1. Humans produce two types of cells: body cells and sex cells.
2. Environmental factors can cause variation among individuals.
3. Two parents always produce the best offspring.
4. Cloning produces identical individuals from one cell.
5. All organisms have two parents.
6. Asexual reproduction occurs only in microorganisms.

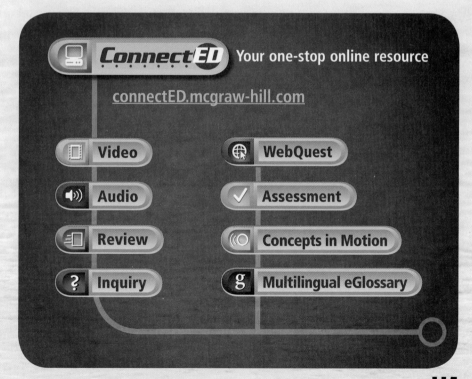

Lesson 1

Sexual Reproduction and Meiosis

Reading Guide

Key Concepts
ESSENTIAL QUESTIONS

- What is sexual reproduction, and why is it beneficial?
- What is the order of the phases of meiosis, and what happens during each phase?
- Why is meiosis important?

Vocabulary

sexual reproduction p. 117
egg p. 117
sperm p. 117
fertilization p. 117
zygote p. 117
diploid p. 118
homologous chromosomes p. 118
haploid p. 119
meiosis p. 119

 Multilingual eGlossary

 Video BrainPOP®

Inquiry Modern Art?

This photo looks like a piece of modern art. It is actually an image of plant cells. The cells are dividing by a process that occurs during the production of sex cells.

Launch Lab

15 minutes

Why do offspring look different?

Unless you're an identical twin, you probably don't look exactly like any siblings you might have. You might have differences in physical characteristics such as eye color, hair color, ear shape, or height. Why are there differences in the offspring from the same parents?

1. Read and complete a lab safety form.
2. Open the **paper bag** labeled *Male Parent*, and, without looking, remove three **beads**. Record the bead colors in your Science Journal, and replace the beads.
3. Open the **paper bag** labeled *Female Parent*, and remove three **beads**. Record the bead colors, and replace the beads.
4. Repeat steps 2 and 3 for each member of the group.
5. After each member has recorded his or her bead colors, study the results. Each combination of male and female beads represents an offspring.

Think About This

1. Compare your group's offspring to another group's offspring. What similarities or differences do you observe?
2. What caused any differences you observed? Explain.
3. 🔑 **Key Concept** Why might this type of reproduction be beneficial to an organism?

What is sexual reproduction?

Have you ever seen a litter of kittens? One kitten might have orange fur like its mother. A second kitten might have gray fur like its father. Still another kitten might look like a combination of both parents. How is this possible?

The kittens look different because of sexual reproduction. **Sexual reproduction** *is a type of reproduction in which the genetic materials from two different cells combine, producing an offspring.* The cells that combine are called sex cells. Sex cells form in reproductive organs. *The female sex cell, an* **egg**, *forms in an ovary. The male sex cell, a* **sperm**, *forms in a testis. During a process called* **fertilization** (fur tuh luh ZAY shun), *an egg cell and a sperm cell join together.* This produces a new cell. *The new cell that forms from fertilization is called a* **zygote**. As shown in **Figure 1**, the zygote develops into a new organism.

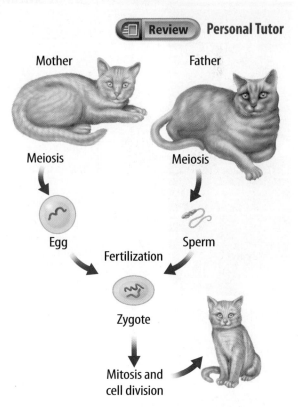

Figure 1 The zygote that forms during fertilization can become a multicellular organism.

Diploid Cells

Following fertilization, a zygote goes through mitosis and cell division. These processes produce nearly all the cells in a multicellular organism. Organisms that reproduce sexually form two kinds of cells—body cells and sex cells. In body cells of most organisms, similar chromosomes occur in pairs. **Diploid** *cells are cells that have pairs of chromosomes.*

Chromosomes

Pairs of chromosomes that have genes for the same traits arranged in the same order are called **homologous** (huh MAH luh gus) **chromosomes.** Because one chromosome is inherited from each parent, the chromosomes are not identical. For example, the kittens mentioned earlier in this lesson inherited a gene for orange fur color from their mother. They also inherited a gene for gray fur color from their father. So, some kittens might be orange, and some might be gray. Both genes for fur color are at the same place on homologous chromosomes, but they code for different colors.

Different organisms have different numbers of chromosomes. Recall that diploid cells have pairs of chromosomes. Notice in **Table 1** that human diploid cells have 23 pairs of chromosomes for a total of 46 chromosomes. A fruit fly diploid cell has 4 pairs of chromosomes, and a rice diploid cell has 12 pairs of chromosomes.

Table 1 An organism's chromosomes can be matched as pairs of chromosomes that have genes for the same traits.

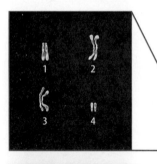

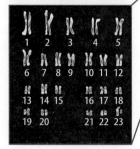

Concepts in Motion Interactive Table

Table 1 Chromosomes of Selected Organisms		
Organism	Number of Chromosomes	Number of Homologous Pairs
Fruit fly	8	4
Rice	24	12
Yeast	32	16
Cat	38	19
Human	46	23
Dog	78	39
Fern	1,260	630

Having the correct number of chromosomes is very important. If a zygote has too many or too few chromosomes, it will not develop properly. For example, a genetic condition called Down syndrome occurs when a person has an extra copy of chromosome 21. A person with Down syndrome can have short stature, heart defects, or mental disabilities.

Haploid Cells

Organisms that reproduce sexually also form egg and sperm cells, or sex cells. Sex cells have only one chromosome from each pair of chromosomes. **Haploid** *cells are cells that have only one chromosome from each pair.* Organisms produce sex cells using a special type of cell division called meiosis. *In* **meiosis**, *one diploid cell divides and makes four haploid sex cells.* Meiosis occurs only during the formation of sex cells.

 Reading Check How do diploid cells differ from haploid cells?

The Phases of Meiosis

Next, you will read about the phases of meiosis. Many of the phases might seem familiar to you because they also occur during mitosis. Recall that mitosis and cytokinesis involve one division of the nucleus and the cytoplasm. Meiosis involves two divisions of the nucleus and the cytoplasm. These divisions are called meiosis I and meiosis II. They result in four haploid cells—cells with half the number of chromosomes as the original cell. As you read about meiosis, think about how it produces sex cells with a reduced number of chromosomes.

WORD ORIGIN

haploid
from Greek *haploeides*, means "single"

FOLDABLES

Make a shutter-fold book and label it as shown. Use it to describe and illustrate the phases of meiosis.

Inquiry MiniLab
20 minutes

How does one cell produce four cells?

When a diploid cell goes through meiosis, it produces four haploid cells. How does this happen?

1. Read and complete a lab safety form.
2. Make a copy of the diagram by tracing circles around a **jar lid** on your **paper**. Label as shown.
3. Use **chenille craft wires** to make red and blue duplicated chromosomes 2.5 cm long and green and yellow duplicated chromosomes 1.5 cm long. Recall that a duplicated chromosome has two sister chromatids connected at the centromere.
4. Place the chromosomes in the diploid cell.
5. Move one long chromosome and one short chromosome into each of the middle cells.
6. Separate the two strands of the chromosomes, and place one strand into each of the haploid cells.

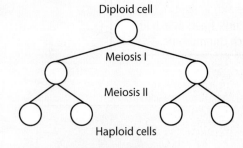

Analyze and Conclude

1. **Describe** What happened to the chromosomes during meiosis I? During meiosis II?
2. **Think Critically** Why are two haploid cells (sperm and egg) needed to form a zygote?
3. **Key Concept** How does one cell form four cells during meiosis?

Lesson 1
EXPLAIN
119

Phases of Meiosis I

A reproductive cell goes through interphase before beginning meiosis I, which is shown in **Figure 2**. During interphase, the reproductive cell grows and copies, or duplicates, its chromosomes. Each duplicated chromosome consists of two sister chromatids joined together by a centromere.

1 Prophase I In the first phase of meiosis I, duplicated chromosomes condense and thicken. Homologous chromosomes come together and form pairs. The membrane surrounding the nucleus breaks apart, and the nucleolus disappears.

2 Metaphase I Homologous chromosome pairs line up along the middle of the cell. A spindle fiber attaches to each chromosome.

3 Anaphase I Chromosome pairs separate and are pulled toward the opposite ends of the cell. Notice that the sister chromatids stay together.

4 Telophase I A nuclear membrane forms around each group of duplicated chromosomes. The cytoplasm divides through cytokinesis and two daughter cells form. Sister chromatids remain together.

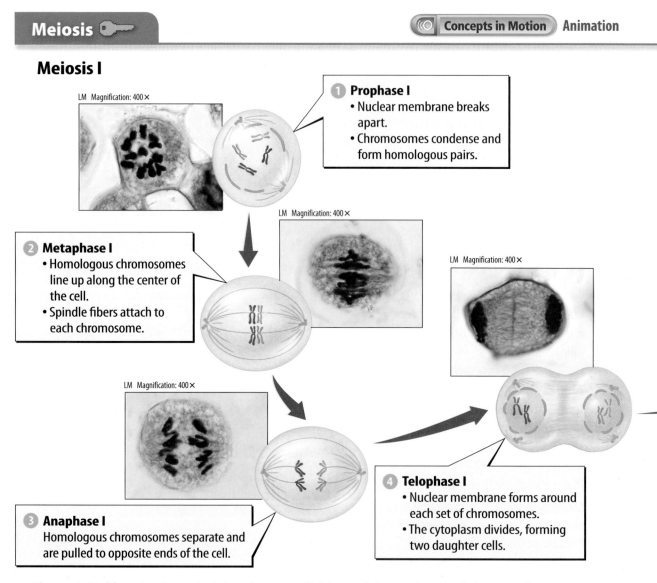

Figure 2 Unlike mitosis, meiosis involves two divisions of the nucleus and the cytoplasm.

Phases of Meiosis II

After meiosis I, the two cells formed during this stage go through a second division of the nucleus and the cytoplasm. This process, shown in **Figure 2**, is called meiosis II.

5 Prophase II Chromosomes are not copied again before prophase II. They remain as condensed, thickened sister chromatids. The nuclear membrane breaks apart, and the nucleolus disappears in each cell.

6 Metaphase II The pairs of sister chromatids line up along the middle of the cell in single file.

7 Anaphase II The sister chromatids of each duplicated chromosome are pulled away from each other and move toward opposite ends of the cells.

8 Telophase II During the final phase of meiosis—telophase II—a nuclear membrane forms around each set of chromatids, which are again called chromosomes. The cytoplasm divides through cytokinesis, and four haploid cells form.

 Key Concept Check List the phases of meiosis in order.

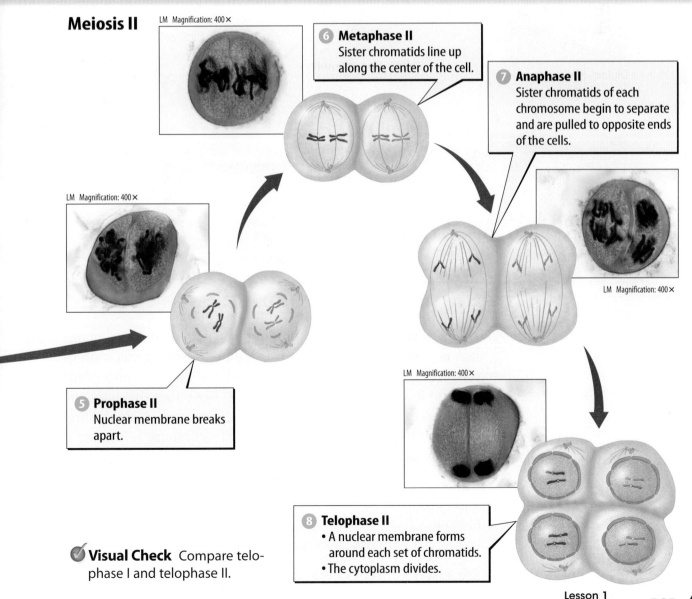

Meiosis II

6 Metaphase II Sister chromatids line up along the center of the cell.

7 Anaphase II Sister chromatids of each chromosome begin to separate and are pulled to opposite ends of the cells.

5 Prophase II Nuclear membrane breaks apart.

8 Telophase II
• A nuclear membrane forms around each set of chromatids.
• The cytoplasm divides.

Visual Check Compare telophase I and telophase II.

Why is meiosis important?

Meiosis forms sex cells with the correct haploid number of chromosomes. This maintains the correct diploid number of chromosomes in organisms when sex cells join. Meiosis also creates genetic variation by producing haploid cells.

Maintaining Diploid Cells

Recall that diploid cells have pairs of chromosomes. Meiosis helps to maintain diploid cells in offspring by making haploid sex cells. When haploid sex cells join together during fertilization, they make a diploid zygote, or fertilized egg. The zygote then divides by mitosis and cell division and creates a diploid organism. **Figure 3** illustrates how the diploid number is maintained in ducks.

Figure 3 Meiosis ensures that the chromosome number of a species stays the same from generation to generation.

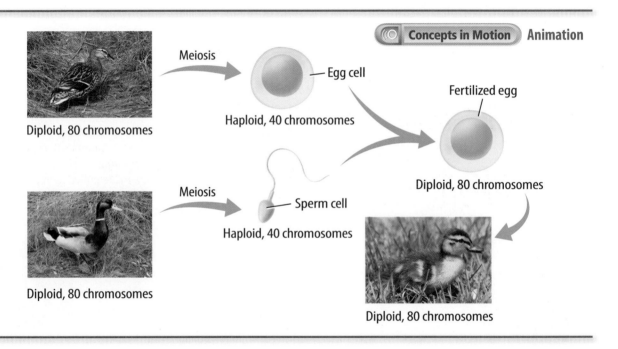

Creating Haploid Cells

The result of meiosis is haploid sex cells. This helps maintain the correct number of chromosomes in each generation of offspring. The formation of haploid cells also is important because it allows for genetic variation. How does this happen? Sex cells can have different sets of chromosomes, depending on how chromosomes line up during metaphase I. Because a cell only gets one chromosome from each pair of homologous chromosomes, the resulting sex cells can be different.

The genetic makeup of offspring is a combination of chromosomes from two sex cells. Variation in the sex cells results in more genetic variation in the next generation.

Key Concept Check Why is meiosis important?

How do mitosis and meiosis differ?

Sometimes, it's hard to remember the differences between mitosis and meiosis. Use **Table 2** to review these processes.

During mitosis and cell division, a body cell and its nucleus divide once and produce two identical cells. These processes are important for growth and repair or replacement of damaged tissue. Some organisms reproduce by these processes. The two daughter cells produced by mitosis and cell division have the same genetic information.

During meiosis, a reproductive cell and its nucleus divide twice and produce four cells—two pairs of identical haploid cells. Each cell has half the number of chromosomes as the original cell. Meiosis happens in the reproductive organs of multicellular organisms. Meiosis forms sex cells used for sexual reproduction.

 Reading Check How many cells are produced during mitosis? During meiosis?

Table 2 Comparison of Types of Cell Division

Characteristic	Meiosis	Mitosis and Cell Division
Number of chromosomes in parent cell	diploid	diploid
Type of parent cell	reproductive	body
Number of divisions of nucleus	2	1
Number of daughter cells produced	4	2
Chromosome number in daughter cells	haploid	diploid
Function	forms sperm and egg cells	growth, cell repair, some types of reproduction

Math Skills

Use Proportions

An equation that shows that two ratios are equivalent is a proportion. The ratios $\frac{1}{2}$ and $\frac{3}{6}$ are equivalent, so they can be written as $\frac{1}{2} = \frac{3}{6}$.

You can use proportions to figure out how many daughter cells will be produced during mitosis. If you know that one cell produces two daughter cells at the end of mitosis, you can use proportions to calculate how many daughter cells will be produced by eight cells undergoing mitosis.

Set up an equation of the two ratios. $\frac{1}{2} = \frac{8}{y}$

Cross-multiply. $1 \times y = 8 \times 2$

$1y = 16$

Divide each side by 1. $y = 16$

Practice

You know that one cell produces four daughter cells at the end of meiosis. How many daughter cells would be produced if eight sex cells undergo meiosis?

- Math Practice
- Personal Tutor

Advantages of Sexual Reproduction

Did you ever wonder why a brother and a sister might not look alike? The answer is sexual reproduction. The main advantage of sexual reproduction is that offspring inherit half their DNA from each parent. Offspring are not likely to inherit the same DNA from the same parents. Different DNA means that each offspring has a different set of traits. This results in genetic variation among the offspring.

Key Concept Check Why is sexual reproduction beneficial?

Genetic Variation

As you just read, genetic variation exists among humans. You can look at your friends to see genetic variation. Genetic variation occurs in all organisms that reproduce sexually. Consider the plants shown in **Figure 4.** The plants are members of the same species, but they have different traits, such as the ability to resist disease.

Due to genetic variation, individuals within a population have slight differences. These differences might be an advantage if the environment changes. Some individuals might have traits that enable them to survive unusually harsh conditions such as a drought or severe cold. Other individuals might have traits that make them resistant to disease.

REVIEW VOCABULARY
DNA
the genetic information in a cell

Genetic Variation

Disease-resistant cassava leaves

Cassava leaves with cassava mosaic disease

Figure 4 These plants belong to the same species. However, one is more disease-resistant than the other.

Visual Check How does cassava mosaic disease affect cassava leaves?

Selective Breeding

Did you know that broccoli, kohlrabi, kale, and cabbage all descended from one type of mustard plant? It's true. More than 2,000 years ago farmers noticed that some mustard plants had different traits, such as larger leaves or bigger flower buds. The farmers started to choose which traits they wanted by selecting certain plants to reproduce and grow. For example, some farmers chose only the plants with the biggest flowers and stems and planted their seeds. Over time, the offspring of these plants became what we know today as broccoli, shown in **Figure 5**. This process is called selective breeding. Selective breeding has been used to develop many types of plants and animals with desirable traits. It is another example of the benefits of sexual reproduction.

Figure 5 The wild mustard is the common ancestor to all these plants.

Disadvantages of Sexual Reproduction

Although sexual reproduction produces more genetic variation, it does have some disadvantages. Sexual reproduction takes time and energy. Organisms have to grow and develop until they are mature enough to produce sex cells. Then the organisms have to form sex cells—either eggs or sperm. Before they can reproduce, organisms usually have to find mates. Searching for a mate can take a long time and requires energy. The search for a mate might also expose individuals to predators, diseases, or harsh environmental conditions. In addition, sexual reproduction is limited by certain factors. For example, fertilization cannot take place during pregnancy, which can last as long as two years in some mammals.

Reading Check What are the disadvantages of sexual reproduction?

Lesson 1 Review

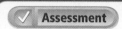

Visual Summary

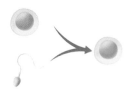

Fertilization occurs when an egg cell and a sperm cell join together.

Organisms produce sex cells through meiosis.

Sexual reproduction results in genetic variation among individuals.

FOLDABLES

Use your lesson Foldable to review the lesson. Save your Foldable for the project at the end of the chapter.

Use Vocabulary

1. **Use the terms** *egg, sperm,* and *zygote* in a sentence.
2. **Distinguish** between haploid and diploid.
3. **Define** *homologous chromosomes* in your own words.

Understand Key Concepts

4. **Define** sexual reproduction.
5. **Draw and label** the phases of meiosis.
6. Homologous chromosomes separate during which phase of meiosis?
   A. anaphase I   C. metaphase I
   B. anaphase II   D. metaphase II

Interpret Graphics

7. **Organize** Copy and fill in the graphic organizer below to sequence the phases of meiosis I and meiosis II.

Critical Thinking

8. **Analyze** Why is the result of the stage of meiosis shown below an advantage for organisms that reproduce sexually?

Math Skills

— Math Practice —

9. If 15 cells undergo meiosis, how many daughter cells would be produced?
10. If each daughter cell from question 9 undergoes meiosis, how many total daughter cells will there be?

What do you think NOW?

You first read the statements below at the beginning of the chapter.

1. Humans produce two types of cells: body cells and sex cells.
2. Environmental factors can cause variation among individuals.
3. Two parents always produce the best offspring.

Did you change your mind about whether you agree or disagree with the statements? Rewrite any false statements to make them true.

AMERICAN MUSEUM OF NATURAL HISTORY

CAREERS in SCIENCE

The Spider Mating Dance

Meet Norman Platnick, a scientist studying spiders.

Norman Platnick is fascinated by all spider species—from the dwarf tarantula-like spiders of Panama to the blind spiders of New Zealand. These are just two of the over 1,400 species he's discovered worldwide.

How does Platnick identify new species? One way is the pedipalps. Every spider has two pedipalps, but they vary in shape and size among the over 40,000 species. Pedipalps look like legs but function more like antennae and mouthparts. Male spiders use their pedipalps to aid in reproduction.

Getting Ready When a male spider is ready to mate, he places a drop of sperm onto a sheet of silk he constructs. Then he dips his pedipalps into the drop to draw up the sperm.

Finding a Mate The male finds a female of the same species by touch or by sensing certain chemicals she releases.

Courting and Mating Males of some species court a female with a special dance. For other species, a male might present a female with a gift, such as a fly wrapped in silk. During mating, the male uses his pedipalps to transfer sperm to the female.

What happens to the male after mating? That depends on the species. Some are eaten by the female, while others move on to find new mates.

▲ Spiders reproduce sexually, so each offspring has a unique combination of genes from its parents. Over many generations, this genetic variation has led to the incredible diversity of spiders in the world today.

◀ Norman Platnick is an arachnologist (uh rak NAH luh just) at the American Museum of Natural History. Arachnologists are scientists who study spiders.

It's Your Turn

RESEARCH Select a species of spider and research its mating rituals. What does a male do to court a female? What is the role of the female? What happens to the spiderlings after they hatch? Use images to illustrate a report on your research.

Lesson 1 EXTEND

Lesson 2

Reading Guide

Key Concepts 🔑
ESSENTIAL QUESTIONS

- What is asexual reproduction, and why is it beneficial?
- How do the types of asexual reproduction differ?

Vocabulary

asexual reproduction p. 129
fission p. 130
budding p. 131
regeneration p. 132
vegetative reproduction p. 133
cloning p. 134

g Multilingual eGlossary

Asexual Reproduction

Inquiry Plants on Plants?

Look closely at the edges of this plant's leaves. Tiny plants are growing there. This type of plant can reproduce without meiosis and fertilization.

Inquiry Launch Lab

20 minutes

How do yeast reproduce?

Some organisms can produce offspring without meiosis or fertilization. You can observe this process when you add sugar and warm water to dried yeast.

1. Read and complete a lab safety form.
2. Pour 125 mL of water into a **beaker.** The water should be at a temperature of 34°C.
3. Add 5 g of **sugar** and 5 g of **yeast** to the water. Stir slightly. Record your observations after 5 minutes in your Science Journal.
4. Using a **dropper,** put a drop of the yeast solution on a **microscope slide.** Place a **coverslip** over the drop.
5. View the yeast solution under a **microscope.** Draw what you see in your Science Journal.

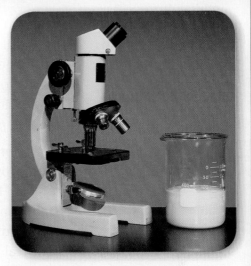

Think About This

1. What evidence did you observe that yeast reproduce?
2. 🔑 **Key Concept** How do you think this process differs from sexual reproduction?

What is asexual reproduction?

Lunch is over and you are in a rush to get to class. You wrap up your half-eaten sandwich and toss it into your locker. A week goes by before you spot the sandwich in the corner of your locker. The surface of the bread is now covered with fuzzy mold—not very appetizing. How did that happen?

The mold on the sandwich is a type of fungus (FUN gus). A fungus releases enzymes that break down organic matter, such as food. It has structures that penetrate and anchor to food, much like roots anchor plants to soil. A fungus can multiply quickly in part because generally a fungus can reproduce either sexually or asexually. Recall that sexual reproduction involves two parent organisms and the processes of meiosis and fertilization. Offspring inherit half their DNA from each parent, resulting in genetic variation among the offspring.

In **asexual reproduction,** *one parent organism produces offspring without meiosis and fertilization.* Because the offspring inherit all their DNA from one parent, they are genetically identical to each other and to their parent.

🔑 **Key Concept Check** Describe asexual reproduction in your own words.

FOLDABLES

Fold a sheet of paper into a six-celled chart. Label the front "Asexual Reproduction," and label the chart inside as shown. Use it to compare types of asexual reproduction.

Fission	Mitotic cell division	Budding
Animal regeneration	Vegetative reproduction	Cloning

Lesson 2
EXPLORE

Types of Asexual Reproduction

There are many different types of organisms that reproduce by asexual reproduction. In addition to fungi, bacteria, protists, plants, and animals can reproduce asexually. In this lesson, you will learn how organisms reproduce asexually.

Fission

Recall that prokaryotes have a simpler cell structure than eukaryotes. A prokaryote's DNA is not contained in a nucleus. For this reason, mitosis does not occur and cell division in a prokaryote is a simpler process than in a eukaryote. *Cell division in prokaryotes that forms two genetically identical cells is known as* **fission.**

Fission begins when a prokaryote's DNA molecule is copied. Each copy attaches to the cell membrane. Then the cell begins to grow longer, pulling the two copies of DNA apart. At the same time, the cell membrane begins to pinch inward along the middle of the cell. Finally the cell splits and forms two new identical offspring. The original cell no longer exists.

As shown in **Figure 6,** *E. coli,* a common bacterium, divides through fission. Some bacteria can divide every 20 minutes. At that rate, 512 bacteria can be produced from one original bacterium in about three hours.

Reading Check What advantage might asexual reproduction by fission have over sexual reproduction?

> **WORD ORIGIN**
> **fission**
> from Latin *fissionem*, means "a breaking up, cleaving"

Fission

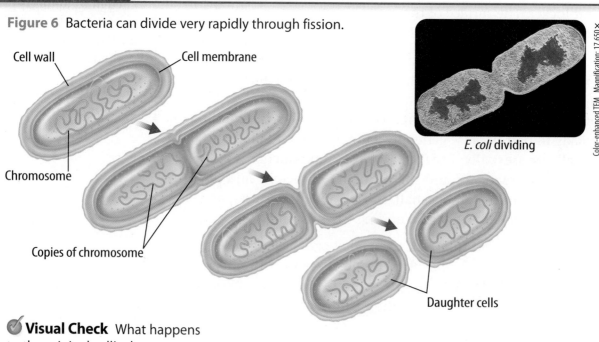

Figure 6 Bacteria can divide very rapidly through fission.

Cell wall — Cell membrane — Chromosome — Copies of chromosome — Daughter cells

E. coli dividing
Color-enhanced TEM Magnification: 17,650×

Visual Check What happens to the original cell's chromosome during fission?

Mitotic Cell Division

Many unicellular eukaryotes reproduce by mitotic cell division. In this type of asexual reproduction, an organism forms two offspring through mitosis and cell division. In **Figure 7**, an amoeba's nucleus has divided by mitosis. Next, the cytoplasm and its contents divide through cytokinesis and two new amoebas form.

Budding

In **budding**, *a new organism grows by mitosis and cell division on the body of its parent.* The bud, or offspring, is genetically identical to its parent. When the bud becomes large enough, it can break from the parent and live on its own. In some cases, an offspring remains attached to its parent and starts to form a colony. **Figure 8** shows a hydra in the process of budding. The hydra is an example of a multicellular organism that can reproduce asexually. Unicellular eukaryotes, such as yeast, can also reproduce through budding, as you saw in the Launch Lab.

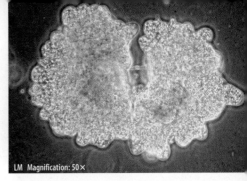

▲ **Figure 7** During mitotic cell division, an amoeba divides its chromosomes and cell contents evenly between the daughter cells.

Budding

Figure 8 The hydra bud has the same genetic makeup as its parent.

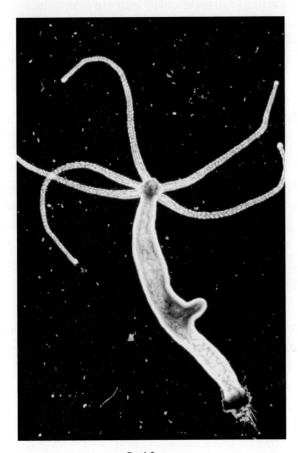

Bud forms.

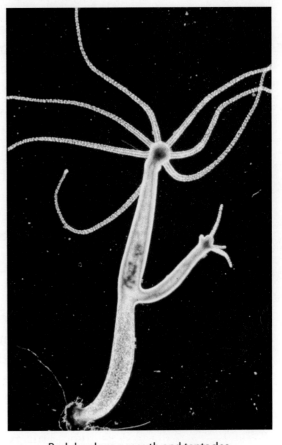

Bud develops a mouth and tentacles.

Animal Regeneration

Figure 9 A planarian can reproduce through regeneration.

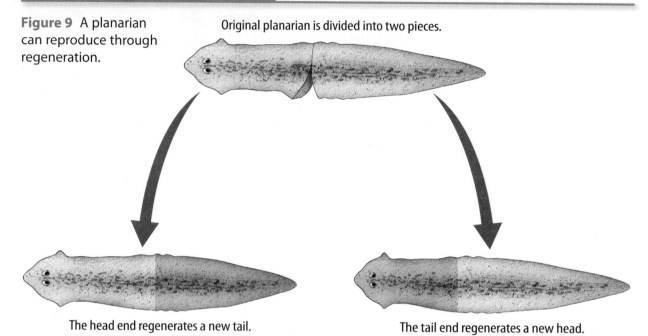

Original planarian is divided into two pieces.

The head end regenerates a new tail.

The tail end regenerates a new head.

Animal Regeneration

*Another type of asexual reproduction, **regeneration**, occurs when an offspring grows from a piece of its parent.* The ability to regenerate a new organism varies greatly among animals.

Producing New Organisms Some sea stars have five arms. If separated from the parent sea star, each arm has the potential to grow into a new organism. To regenerate a new sea star, the arm must contain a part of the central disk of the parent. If conditions are right, one five-armed sea star can produce as many as five new organisms.

Sea urchins, sea cucumbers, sponges, and planarians, such as the one shown in **Figure 9,** can also reproduce through regeneration. Notice that each piece of the original planarian becomes a new organism. As with all types of asexual reproduction, the offspring is genetically identical to the parent.

Reading Check What is true of all cases of asexual reproduction?

Producing New Parts When you hear the term *regeneration*, you might think about a salamander regrowing a lost tail or leg. Regeneration of damaged or lost body parts is common in many animals. Newts, tadpoles, crabs, hydra, and zebra fish are all able to regenerate body parts. Even humans are able to regenerate some damaged body parts, such as the skin and the liver. This type of regeneration, however, is not considered asexual reproduction. It does not produce a new organism.

ACADEMIC VOCABULARY
potential
(noun) possibility

Vegetative Reproduction

Plants can also reproduce asexually in a process similar to regeneration. **Vegetative reproduction** *is a form of asexual reproduction in which offspring grow from a part of a parent plant.* For example, the strawberry plants shown in **Figure 10** send out long horizontal stems called stolons. Wherever a stolon touches the ground, it can produce roots. Once the stolons have grown roots, a new plant can grow—even if the stolons have broken off the parent plant. Each new plant grown from a stolon is genetically identical to the parent plant.

Vegetative reproduction usually involves structures such as the roots, the stems, and the leaves of plants. In addition to strawberries, many other plants can reproduce by this method, including raspberries, potatoes, and geraniums.

Figure 10 The smaller plants were grown from stolons produced by the parent plant.

Visual Check Which plants in the figure are the parent plants?

Inquiry MiniLab

15 minutes

What parts of plants can grow?

You probably know that plants can grow from seeds. But you might be surprised to learn that other parts of plants can grow and produce a new plant.

1. Carefully examine the photos of vegetative reproduction.
2. Create a data chart in your Science Journal to record your observations. Identify which part of the plant (leaf, stem, etc.) would be used to grow a new plant.

Analyze and Conclude

1. **Explain** How is the vegetative reproduction you observed a kind of asexual reproduction?
2. **Infer** how farmers or gardeners might use vegetative reproduction.
3. **Key Concept** Describe a method you might use to produce a new plant using vegetative reproduction.

Lesson 2
EXPLAIN

Cloning

Fission, budding, and regeneration are all types of asexual reproduction that can produce genetically identical offspring in nature. In the past, the term *cloning* described any process that produced genetically identical offspring. Today, however, the word usually refers to a technique developed by scientists and performed in laboratories. **Cloning** *is a type of asexual reproduction performed in a laboratory that produces identical individuals from a cell or from a cluster of cells taken from a multicellular organism.* Farmers and scientists often use cloning to make copies of organisms or cells that have desirable traits, such as large flowers.

Plant Cloning Some plants can be cloned using a method called tissue culture, as shown in **Figure 11.** Tissue culture enables plant growers and scientists to make many copies of a plant with desirable traits, such as sweet fruit. Also, a greater number of plants can be produced more quickly than by vegetative reproduction.

Tissue culture also enables plant growers to reproduce plants that might have become infected with a disease. To clone such a plant, a scientist can use cells from a part of a plant where they are rapidly undergoing mitosis and cell division. This part of a plant is called a meristem. Cells in meristems are disease-free. Therefore, if a plant becomes infected with a disease, it can be cloned using meristem cells.

SCIENCE USE V. COMMON USE

culture
Science Use the process of growing living tissue in a laboratory

Common Use the social customs of a group of people

Figure 11 New carrot plants can be produced from cells of a carrot root using tissue culture techniques.

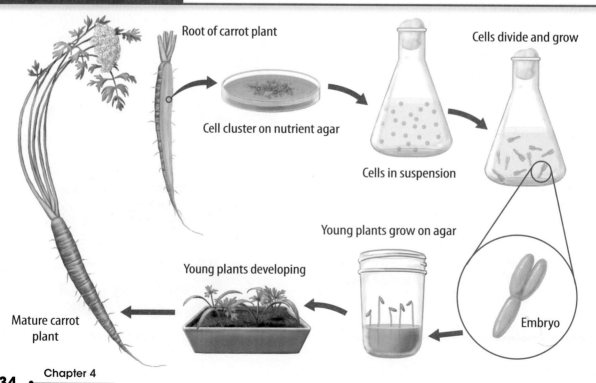

Plant Cloning

Animal Cloning In addition to cloning plants, scientists have been able to clone many animals. Because all of a clone's chromosomes come from one parent (the donor of the nucleus), the clone is a genetic copy of its parent. The first mammal cloned was a sheep named Dolly. **Figure 12** illustrates how this was done.

Scientists are currently working to save some endangered species from extinction by cloning. Although cloning is an exciting advancement in science, some people are concerned about the high cost and the ethics of this technique. Ethical issues include the possibility of human cloning. You might be asked to consider issues like this during your lifetime.

 Key Concept Check Compare and contrast the different types of asexual reproduction.

Figure 12 Scientists used two different sheep to produce the cloned sheep known as Dolly.

Animal Cloning

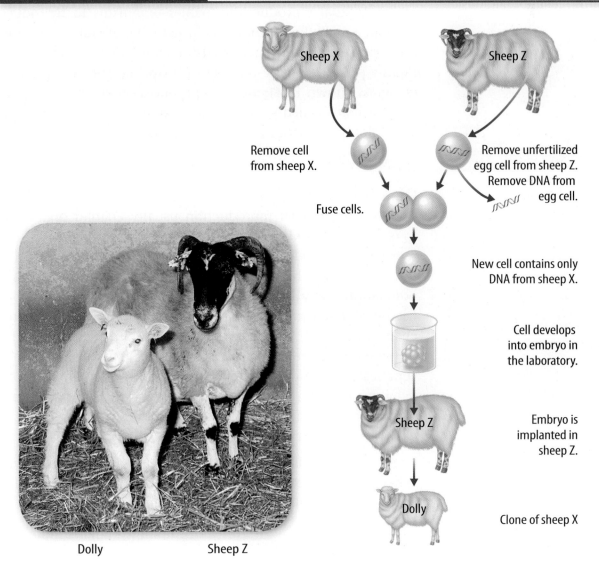

Dolly   Sheep Z

Lesson 2
EXPLAIN
135

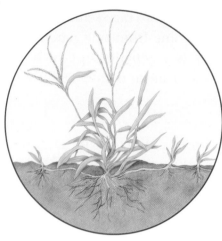

Figure 13 Crabgrass can spread quickly because it reproduces asexually.

Advantages of Asexual Reproduction

What are the advantages to organisms of reproducing asexually? Asexual reproduction enables organisms to reproduce without a mate. Recall that searching for a mate takes time and energy. Asexual reproduction also enables some organisms to rapidly produce a large number of offspring. For example, the crabgrass shown in **Figure 13** reproduces asexually by underground stems called stolons. This enables one plant to spread and colonize an area in a short period of time.

 Key Concept Check How is asexual reproduction beneficial?

Disadvantages of Asexual Reproduction

Although asexual reproduction usually enables organisms to reproduce quickly, it does have some disadvantages. Asexual reproduction produces offspring that are genetically identical to their parent. This results in little genetic variation within a population. Why is genetic variation important? Recall from Lesson 1 that genetic variation can give organisms a better chance of surviving if the environment changes. Think of the crabgrass. Imagine that all the crabgrass plants in a lawn are genetically identical to their parent plant. If a certain weed killer can kill the parent plant, then it can kill all the crabgrass plants in the lawn. This might be good for your lawn, but it is a disadvantage for the crabgrass.

Another disadvantage of asexual reproduction involves genetic changes, called mutations, that can occur. If an organism has a harmful mutation in its cells, the mutation will be passed to asexually reproduced offspring. This could affect the offspring's ability to survive.

Lesson 2 Review

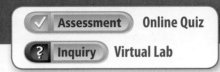

Visual Summary

In asexual reproduction, offspring are produced without meiosis and fertilization.

Cloning is one type of asexual reproduction.

Asexual reproduction enables organisms to reproduce quickly.

FOLDABLES

Use your lesson Foldable to review the lesson. Save your Foldable for the project at the end of the chapter.

What do you think NOW?

You first read the statements below at the beginning of the chapter.

4. Cloning produces identical individuals from one cell.

5. All organisms have two parents.

6. Asexual reproduction occurs only in microorganisms.

Did you change your mind about whether you agree or disagree with the statements? Rewrite any false statements to make them true.

Use Vocabulary

1 In _____ _____, only one parent organism produces offspring.

2 Define the term *cloning* in your own words.

3 Use the term *regeneration* in a sentence.

Understand Key Concepts

4 State two reasons why asexual reproduction is beneficial.

5 Which is an example of asexual reproduction by regeneration?
   A. cloning sheep
   B. lizard regrowing a tail
   C. sea star arm producing a new organism
   D. strawberry plant producing stolons

6 Construct a chart that includes an example of each type of asexual reproduction.

Interpret Graphics

7 Examine the diagram below and write a short paragraph describing the process of tissue culture.

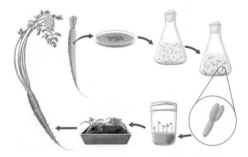

8 Organize Copy and fill in the graphic organizer below to list the different types of asexual reproduction that occur in multicellular organisms.

Critical Thinking

9 Justify the use of cloning to save endangered animals.

Lesson 2
EVALUATE
137

Inquiry Lab

40 minutes

Mitosis and Meiosis

Materials

pool noodles

Safety

During cellular reproduction, many changes occur in the nucleus of cells involving the chromosomes. You could think about these changes as a set of choreographed moves like you would see in a dance. In this lab you will act out the moves that chromosomes make during mitosis and meiosis in order to understand the steps that occur when cells reproduce.

Ask a Question
How do chromosomes change and move during mitosis and meiosis?

Make Observations

1. Read and complete a lab safety form.
2. Form a cell nucleus with four chromosomes represented by students holding four different colors of pool noodles. Other students play the part of the nuclear membrane and form a circle around the chromosomes.

3. The chromosomes duplicate during interphase. Each chromosome is copied, creating a chromosome with two sister chromatids.
4. Perform mitosis.
   a. During prophase, the nuclear membrane breaks apart, and the nucleolus disappears.
   b. In metaphase, duplicated chromosomes align in the middle of the cell.
   c. The sister chromatids separate in anaphase.
   d. In telophase, the nuclear membrane reforms around two daughter cells.
5. Repeat steps 2 and 3. Perform meiosis.
   a. In prophase I, the nuclear membrane breaks apart, the nucleolus disappears, and homologous chromosomes pair up.
   b. In metaphase I, homologous chromosomes line up along the center of the cell.
   c. During anaphase I, the pairs of homologous chromosomes separate.
   d. In telophase I, the nuclear membrane reforms.
   e. Each daughter cell now performs meiosis II independently. In prophase II, the nuclear membrane breaks down, and the nucleolus disappears.
   f. During metaphase II, duplicated chromosomes align in the middle of the cell.

Chapter 4
EXTEND

g. Sister chromatids separate in anaphase II.

h. In telophase II, the nuclear membrane reforms.

Form a Hypothesis

6 Use your observations to form a hypothesis about the results of an error in meiosis. For example, you might explain the results of an error during anaphase I.

Test your Hypothesis

7 Perform meiosis, incorporating the error you chose in step 6.

8 Compare the outcome to your hypothesis. Does your data support your hypothesis? If not, revise your hypothesis and repeat steps 6–8.

Analyze and Conclude

9 **Compare and Contrast** How are mitosis and meiosis I similar? How are they different?

10 **The Big Idea** What is the difference between the chromosomes in cells at the beginning and the end of mitosis? At the beginning and end of meiosis?

11 **Critique** How did performing cellular replications using pool noodles help you understand mitosis and meiosis?

Communicate Your Results

Create a chart of the changes and movements of chromosomes in each of the steps in meiosis and mitosis. Include colored drawings of chromosomes and remember to draw the cell membranes.

Investigate some abnormalities that occur when mistakes are made during mitosis or meiosis. Draw a chart of the steps of reproduction showing how the mistake is made. Write a short description of the problems that result from the mistake.

Lab Tips

☑ Figure out where the boundaries of your cell are before you start.

☑ Review the phases of mitosis and meiosis before beginning to act out how the chromosomes move during each process.

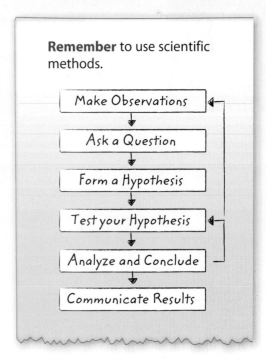

Remember to use scientific methods.

- Make Observations
- Ask a Question
- Form a Hypothesis
- Test your Hypothesis
- Analyze and Conclude
- Communicate Results

Chapter 4 Study Guide

Reproduction ensures the survival of species.

Key Concepts Summary

Lesson 1: Sexual Reproduction and Meiosis

- **Sexual reproduction** is the production of an offspring from the joining of a **sperm** and an **egg**.
- Division of the nucleus and cytokinesis happens twice in **meiosis**. Meiosis I separates homologous chromosomes. Meiosis II separates sister chromatids.
- Meiosis maintains the chromosome number of a species from one generation to the next.

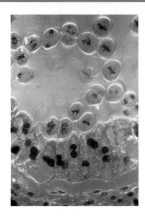

Vocabulary

sexual reproduction p. 117
egg p. 117
sperm p. 117
fertilization p. 117
zygote p. 117
diploid p. 118
homologous chromosomes p. 118
haploid p. 119
meiosis p. 119

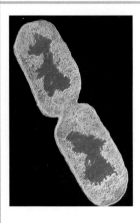

Lesson 2: Asexual Reproduction

- **Asexual reproduction** is the production of offspring by one parent, which results in offspring that are genetically identical to the parent.
- Types of asexual reproduction include **fission**, mitotic cell division, **budding**, **regeneration**, **vegetative reproduction**, and **cloning**.
- Asexual reproduction can produce a large number of offspring in a short amount of time.

asexual reproduction p. 129
fission p. 130
budding p. 131
regeneration p. 132
vegetative reproduction p. 133
cloning p. 134

Study Guide

- Personal Tutor
- Vocabulary eGames
- Vocabulary eFlashcards

Chapter Project

Assemble your lesson Foldables as shown to make a Chapter Project. Use the project to review what you have learned in this chapter.

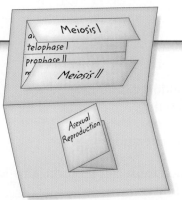

Use Vocabulary

1. Define *meiosis* in your own words.
2. Distinguish between an *egg* and a *zygote*.
3. Use the vocabulary words *haploid* and *diploid* in a sentence.
4. Cell division in prokaryotes is called _____.
5. Define the term *vegetative reproduction* in your own words.
6. Distinguish between *regeneration* and *budding*.
7. A type of reproduction in which the genetic materials from two different cells combine, producing an offspring, is called _____ _____.

Link Vocabulary and Key Concepts

Concepts in Motion — Interactive Concept Map

Copy this concept map, and then use vocabulary terms from the previous page to complete the concept map.

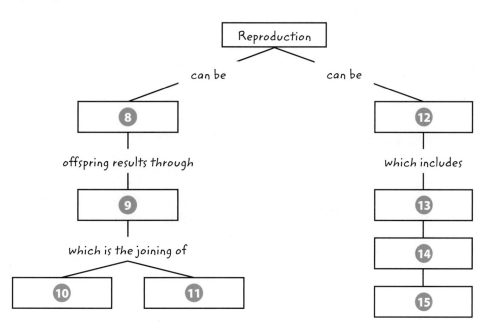

Chapter 4 Review

Understand Key Concepts

1. Which is an advantage of sexual reproduction?
   A. Offspring are identical to the parents.
   B. Offspring with genetic variation are produced.
   C. Organisms don't have to search for a mate.
   D. Reproduction is rapid.

2. Which describes cells that have only one copy of each chromosome?
   A. diploid
   B. haploid
   C. homologous
   D. zygote

Use the figure below to answer questions 3 and 4.

3. Which phase of meiosis I is shown in the diagram?
   A. anaphase I
   B. metaphase I
   C. prophase I
   D. telophase I

4. Which phase of meiosis I comes after the phase in the diagram?
   A. anaphase I
   B. metaphase I
   C. prophase I
   D. telophase I

5. Tissue culture is an example of which type of reproduction?
   A. budding
   B. cloning
   C. fission
   D. regeneration

6. Which type of asexual reproduction is shown in the figure below?

   A. budding
   B. cloning
   C. fission
   D. regeneration

7. A bacterium can reproduce by which method?
   A. budding
   B. cloning
   C. fission
   D. regeneration

8. Which statement best describes why genetic variation is beneficial to populations of organisms?
   A. Individuals look different from one another.
   B. Only one parent is needed to produce offspring.
   C. Populations of the organism increase more rapidly.
   D. Species can better survive environmental changes.

9. In which phase of meiosis II do sister chromatids line up along the center of the cell?
   A. anaphase II
   B. metaphase II
   C. prophase II
   D. telophase II

Chapter Review

Critical Thinking

10. **Contrast** haploid cells and diploid cells.

11. **Model** Make a model of homologous chromosomes using materials of your choice.

12. **Form a hypothesis** about the effect of a mistake in separating homologous chromosomes during meiosis.

13. **Analyze** Crabgrass reproduces asexually by vegetative reproduction. Use the figure below to explain why this form of reproduction is an advantage for the crabgrass.

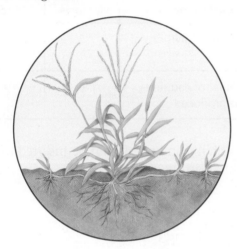

14. **Compare** budding and cloning.

15. **Create** a table showing the advantages and disadvantages of asexual reproduction.

16. **Compare and contrast** sexual reproduction and asexual reproduction.

Writing in Science

17. **Create** a plot for a short story that describes an environmental change and the importance of genetic variation in helping a species survive that change. Include characters, a setting, a climax, and an ending for your plot.

REVIEW THE BIG IDEA

18. Think of all the advantages of sexual and asexual reproduction. Use these ideas to summarize why organisms reproduce.

19. The baby penguin below has a mother and a father. Do all living things have two parents? Explain.

Math Skills

Use Proportions

20. During mitosis, the original cell produces two daughter cells. How many daughter cells will be produced if 250 mouse cells undergo mitosis?

21. During meiosis, the original reproductive cell produces four daughter cells. How many daughter cells will be produced if 250 mouse reproductive cells undergo meiosis?

22. Two reproductive cells undergo meiosis. Each daughter cell also undergoes meiosis. How many cells are produced when the daughter cells divide?

Standardized Test Practice

Record your answers on the answer sheet provided by your teacher or on a sheet of paper.

Multiple Choice

1. How do sea stars reproduce?
   A cloning
   B fission
   C animal regeneration
   D vegetative reproduction

Use the diagram below to answer questions 2 and 3.

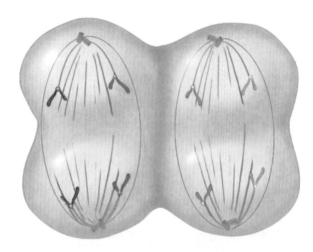

2. What stage of meiosis does the drawing illustrate?
   A anaphase I
   B anaphase II
   C prophase I
   D prophase II

3. Which stage takes place *before* the one in the diagram above?
   A metaphase I
   B metaphase II
   C telophase I
   D telophase II

4. What type of asexual reproduction includes stolons?
   A budding
   B cloning
   C animal regeneration
   D vegetative reproduction

Use the table below to answer question 5.

Comparison of Types of Cell Division		
Characteristic	Meiosis	Mitosis
Number of divisions of nucleus	2	A
Number of daughter cells produced	B	2

5. Which numbers should be inserted for A and B in the chart?
   A A=1 and B=2
   B A=1 and B=4
   C A=2 and B=2
   D A=2 and B=4

6. Which results in genetic variation?
   A cloning
   B fission
   C sexual reproduction
   D vegetative reproduction

7. Which is NOT true of homologous chromosomes?
   A The are identical.
   B They are in pairs.
   C They have genes for the same traits.
   D They have genes that are in the same order.

Standardized Test Practice

Use the figure below to answer question 8.

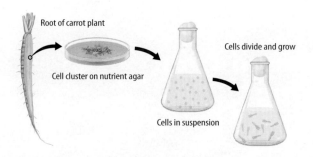

8 The figure illustrates the first four steps of which reproductive process?

A animal cloning

B regeneration

C tissue culture

D vegetative reproduction

9 If 12 reproductive cells undergo meiosis, how many daughter cells will result?

A 12

B 24

C 48

D 60

10 Which is NOT true of asexual reproduction?

A Many offspring can be produced rapidly.

B Offspring are different from the parents.

C Offspring have no genetic variation.

D Organisms can reproduce without a mate.

Constructed Response

Use the figure below to answer questions 11 and 12.

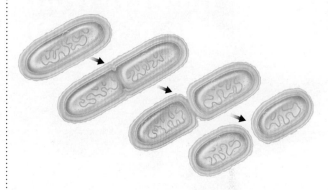

11 Identify the type of asexual reproduction shown in the figure above. How does it differ from sexual reproduction?

12 Compare and contrast budding with the type of asexual reproduction shown in the figure above.

13 What are some differences between the results of selectively breeding plants and cloning them?

14 Use the example of the wild mustard plant to describe the benefits of selective breeding.

15 What are the advantages and disadvantages of cloning animals?

NEED EXTRA HELP?															
If You Missed Question...	1	2	3	4	5	6	7	8	9	10	11	12	13	14	15
Go to Lesson...	2	1	1	2	1	1	1	2	1	2	1,2	2	1,2	1	2

Chapter 5
Genetics

THE BIG IDEA How are traits passed from parents to offspring?

Inquiry How did this happen?

The color of this fawn is caused by a genetic trait called albinism. Albinism is the absence of body pigment. Notice that the fawn's mother has brown fur, the normal fur color of an adult whitetail deer.

- Why do you think the fawn looks so different from its mother?
- What do you think determines the color of the offspring?
- How do you think traits are passed from generation to generation?

Get Ready to Read

What do you think?
Before you read, decide if you agree or disagree with each of these statements. As you read this chapter, see if you change your mind about any of the statements.

1. Like mixing paints, parents' traits always blend in their offspring.
2. If you look more like your mother than you look like your father, then you received more traits from your mother.
3. All inherited traits follow Mendel's patterns of inheritance.
4. Scientists have tools to predict the form of a trait an offspring might inherit.
5. Any condition present at birth is genetic.
6. A change in the sequence of an organism's DNA always changes the organism's traits.

ConnectED Your one-stop online resource

connectED.mcgraw-hill.com

- Video
- WebQuest
- Audio
- Assessment
- Review
- Concepts in Motion
- Inquiry
- Multilingual eGlossary

Lesson 1

Reading Guide

Key Concepts
ESSENTIAL QUESTIONS

- Why did Mendel perform cross-pollination experiments?
- What did Mendel conclude about inherited traits?
- How do dominant and recessive factors interact?

Vocabulary
heredity p. 149
genetics p. 149
dominant trait p. 155
recessive trait p. 155

Multilingual eGlossary

Video BrainPOP®

Mendel and His Peas

Inquiry Same Species?

Have you ever seen a black ladybug? It is less common than the orange variety you might know, but both are the same species of beetle. So why do they look different? Believe it or not, a study of pea plants helped scientists explain these differences.

Launch Lab

10 minutes

What makes you unique?

Traits such as eye color have many different types, but some traits have only two types. By a show of hands, determine how many students in your class have each type of trait below.

Student Traits

Trait	Type 1	Type 2
Earlobes	Unattached	Attached
Thumbs	Curved	Straight
Interlacing fingers	Left thumb over right thumb	Right thumb over left thumb

Think About This

1. Why might some students have types of traits that others do not have?
2. If a person has dimples, do you think his or her offspring will have dimples? Explain.
3. **Key Concept** What do you think determines the types of traits you inherit?

Early Ideas About Heredity

Have you ever mixed two paint colors to make a new color? Long ago, people thought an organism's characteristics, or traits, mixed like colors of paint because offspring resembled both parents. This is known as blending inheritance.

Today, scientists know that **heredity** (huh REH duh tee)—*the passing of traits from parents to offspring*—is more complex. For example, you might have blue eyes but both of your parents have brown eyes. How does this happen? More than 150 years ago, Gregor Mendel, an Austrian monk, performed experiments that helped answer these questions and disprove the idea of blending inheritance. Because of his research, Mendel is known as the father of **genetics** (juh NEH tihks)—*the study of how traits are passed from parents to offspring.*

WORD ORIGIN

genetics
from Greek *genesis*, means "origin"

Lesson 1
EXPLORE

Mendel's Experimental Methods

During the 1850s, Mendel studied genetics by doing controlled breeding experiments with pea plants. Pea plants were ideal for genetic studies because

- they reproduce quickly. This enabled Mendel to grow many plants and collect a lot of data.

- they have easily observed traits, such as flower color and pea shape. This enabled Mendel to observe whether or not a trait was passed from one generation to the next.

- Mendel could control which pairs of plants reproduced. This enabled him to determine which traits came from which plant pairs.

Pollination in Pea Plants

To observe how a trait was inherited, Mendel controlled which plants pollinated other plants. Pollination occurs when pollen lands on the pistil of a flower. **Sperm** cells from the pollen then can fertilize **egg** cells in the pistil. Pollination in pea plants can occur in two ways. Self-pollination occurs when pollen from one plant lands on the pistil of a flower on the same plant, as shown in **Figure 1.** Cross-pollination occurs when pollen from one plant reaches the pistil of a flower on a different plant. Cross-pollination occurs naturally when wind, water, or animals such as bees carry pollen from one flower to another. Mendel allowed one group of flowers to self-pollinate. With another group, he cross-pollinated the plants himself.

REVIEW VOCABULARY

sperm
a haploid sex cell formed in the male reproductive organs

egg
a haploid sex cell formed in the female reproductive organs

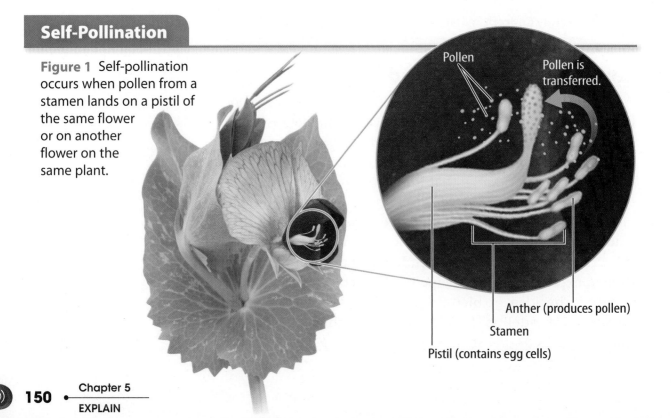

Self-Pollination

Figure 1 Self-pollination occurs when pollen from a stamen lands on a pistil of the same flower or on another flower on the same plant.

True-Breeding Plants

Mendel began his experiments with plants that were true-breeding for the trait he would test. When a true-breeding plant self-pollinates, it always produces offspring with traits that match the parent. For example, when a true-breeding pea plant with wrinkled seeds self-pollinates, it produces only plants with wrinkled seeds. In fact, plants with wrinkled seeds appear generation after generation.

Mendel's Cross-Pollination

By cross-pollinating plants himself, Mendel was able to select which plants pollinated other plants. **Figure 2** shows an example of a manual cross between a plant with white flowers and one with purple flowers.

Figure 2 Mendel removed the stamens of one flower and pollinated that flower with pollen from a flower of a different plant. In this way, he controlled pollination.

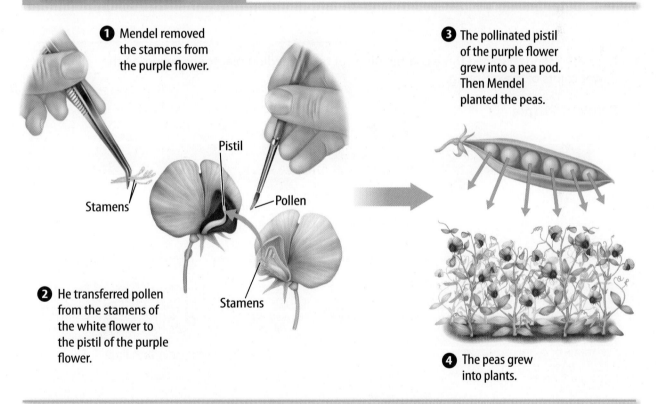

Mendel cross-pollinated hundreds of plants for each set of traits, such as flower color—purple or white; seed color—green or yellow; and seed shape—round or wrinkled. With each cross-pollination, Mendel recorded the traits that appeared in the offspring. By testing such a large number of plants, Mendel was able to predict which crosses would produce which traits.

 Key Concept Check Why did Mendel perform cross-pollination experiments?

Mendel's Results

Once Mendel had enough true-breeding plants for a trait that he wanted to test, he cross-pollinated selected plants. His results are shown in **Figure 3**.

First-Generation Crosses

A cross between true-breeding plants with purple flowers produced plants with only purple flowers. A cross between true-breeding plants with white flowers produced plants with only white flowers. But something unexpected happened when Mendel crossed true-breeding plants with purple flowers and true-breeding plants with white flowers—all the offspring had purple flowers.

New Questions Raised

The results of the crosses between true-breeding plants with purple flowers and true-breeding plants with white flowers led to more questions for Mendel. Why did all the offspring always have purple flowers? Why were there no white flowers? Why didn't the cross produce offspring with pink flowers—a combination of the white and purple flower colors? Mendel carried out more experiments with pea plants to answer these questions.

 Reading Check Predict the offspring of a cross between two true-breeding pea plants with smooth seeds.

First-Generation Crosses

Figure 3 Mendel crossed three combinations of true-breeding plants and recorded the flower colors of the offspring.

Purple × Purple

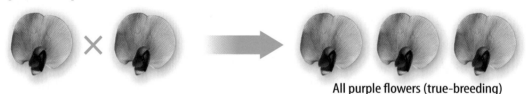

All purple flowers (true-breeding)

White × White

All white flowers (true-breeding)

Purple (true-breeding) × White (true-breeding)

All purple flowers (hybrids)

Visual Check Suppose you cross hundreds of true-breeding plants with purple flowers with hundreds of true-breeding plants with white flowers. Based on the results of this cross in the figure above, would any offspring produce white flowers? Explain.

Second-Generation (Hybrid) Crosses

The first-generation purple-flowering plants are called **hybrid** plants. This means they came from true-breeding parent plants with different forms of the same trait. Mendel wondered what would happen if he cross-pollinated two purple-flowering hybrid plants.

As shown in **Figure 4,** some of the offspring had white flowers, even though both parents had purple flowers. The results were similar each time Mendel cross-pollinated two hybrid plants. The trait that had disappeared in the first generation always reappeared in the second generation.

The same result happened when Mendel cross-pollinated pea plants for other traits. For example, he found that cross-pollinating a true-breeding yellow-seeded pea plant with a true-breeding green-seeded pea plant always produced yellow-seeded hybrids. A second-generation cross of two yellow-seeded hybrids always yielded plants with yellow seeds and plants with green seeds.

Reading Check What is a hybrid plant?

> **SCIENCE USE V. COMMON USE**
> **hybrid**
> **Science Use** the offspring of two animals or plants with different forms of the same trait
>
> **Common Use** having two types of components that perform the same function, such as a vehicle powered by both a gas engine and an electric motor

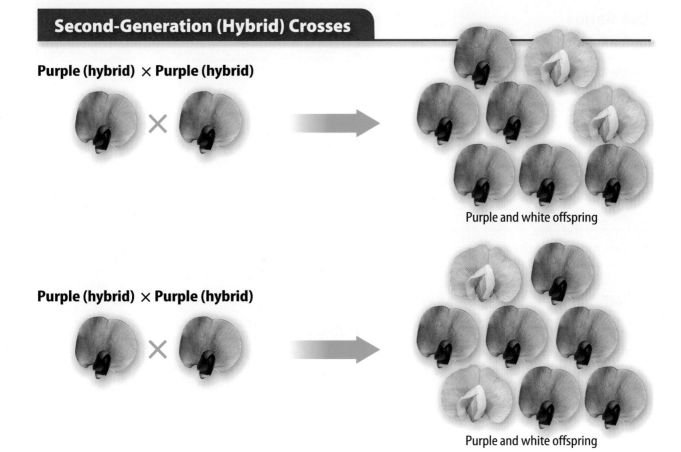

Figure 4 Mendel cross-pollinated first-generation hybrid offspring to produce second-generation offspring. In each case, the trait that had disappeared from the first generation reappeared in the second generation.

Table 1 When Mendel crossed two hybrids for a given trait, the trait that had disappeared then reappeared in a ratio of about 3∶1.

Concepts in Motion Interactive Table

Table 1 Results of Hybrid Crosses

Characteristic	Trait and Number of Offspring	Trait and Number of Offspring	Ratio
Flower color	Purple 705	White 224	3.15∶1
Flower position	Axial (Side of stem) 651	Terminal (End of stem) 207	3.14∶1
Seed color	Yellow 6,022	Green 2,001	3.01∶1
Seed shape	Round 5,474	Wrinkled 1,850	2.96∶1
Pod shape	Inflated (Smooth) 882	Constricted (Bumpy) 299	2.95∶1
Pod color	Green 428	Yellow 152	2.82∶1
Stem length	Long 787	Short 277	2.84∶1

Math Skills

Use Ratios

A ratio is a comparison of two numbers or quantities by division. For example, the ratio comparing 6,022 yellow seeds to 2,001 green seeds can be written as follows:

6,022 to 2,001 or

6,022 : 2,001 or

$\frac{6,022}{2,001}$

To simplify the ratio, divide the first number by the second number.

$\frac{6,022}{2,001} = \frac{3}{1}$ or 3 : 1

Practice
There are 14 girls and 7 boys in a science class. Simplify the ratio.

- Review
- Math Practice
- Personal Tutor

More Hybrid Crosses

Mendel counted and recorded the traits of offspring from many experiments in which he cross-pollinated hybrid plants. Data from these experiments are shown in **Table 1.** He analyzed these data and noticed patterns. For example, from the data of crosses between hybrid plants with purple flowers, he found that the ratio of purple flowers to white flowers was about 3 : 1. This means purple-flowering pea plants grew from this cross three times more often than white-flowering pea plants grew from the cross. He calculated similar ratios for all seven traits he tested.

154 Chapter 5
EXPLAIN

Mendel's Conclusions

After analyzing the results of his experiments, Mendel concluded that two genetic factors control each inherited trait. He also proposed that when organisms reproduce, each reproductive cell—sperm or egg—contributes one factor for each trait.

 Key Concept Check What did Mendel conclude about inherited traits?

Dominant and Recessive Traits

Recall that when Mendel cross-pollinated a true-breeding plant with purple flowers and a true-breeding plant with white flowers, the hybrid offspring had only purple flowers. Mendel hypothesized that the hybrid offspring had one genetic factor for purple flowers and one genetic factor for white flowers. But why were there no white flowers?

Mendel also hypothesized that the purple factor is the only factor seen or expressed because it blocks the white factor. *A genetic factor that blocks another genetic factor is called a* **dominant** (DAH muh nunt) **trait.** A dominant trait, such as purple pea flowers, is observed when offspring have either one or two dominant factors. *A genetic factor that is blocked by the presence of a dominant factor is called a* **recessive** (rih SE sihv) **trait.** A recessive trait, such as white pea flowers, is observed only when two recessive genetic factors are present in offspring.

From Parents to Second Generation

For the second generation, Mendel cross-pollinated two hybrids with purple flowers. About 75 percent of the second-generation plants had purple flowers. These plants had at least one dominant factor. Twenty-five percent of the second-generation plants had white flowers. These plants had the same two recessive factors.

 Key Concept Check How do dominant and recessive factors interact?

FOLDABLES

Make a vertical two-tab book and label it as shown. Use it to organize your notes on dominant and recessive factors.

Inquiry MiniLab   20 minutes

Which is the dominant trait?

Imagine you are Gregor Mendel's lab assistant studying pea plant heredity. Mendel has crossed true-breeding plants with axial flowers and true-breeding plants with terminal flowers. Use the data below to determine which trait is dominant.

Pea Flower Location Results		
Generation	Axial (Number of Offspring)	Terminal (Number of Offspring)
First	794	0
Second	651	207

Analyze and Conclude

1. **Determine** which trait is dominant and which trait is recessive. Support your answer with data.
2. **Key Concept** Analyze the first-generation data. What evidence do you have that one trait is dominant over the other?

Lesson 1 Review

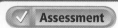 Assessment Online Quiz

Visual Summary

 Genetics is the study of how traits are passed from parents to offspring.

 Mendel studied genetics by doing cross-breeding experiments with pea plants.

 Purple 705
 White 224

Mendel's experiments with pea plants showed that some traits are dominant and others are recessive.

FOLDABLES

Use your lesson Foldable to review the lesson. Save your Foldable for the project at the end of the chapter.

What do you think NOW?

You first read the statements below at the beginning of the chapter.

1. Like mixing paints, parents' traits always blend in their offspring.
2. If you look more like your mother than you look like your father, then you received more traits from your mother.

Did you change your mind about whether you agree or disagree with the statements? Rewrite any false statements to make them true.

Use Vocabulary

1. **Distinguish** between heredity and genetics.
2. **Define** the terms *dominant* and *recessive*.
3. **Use the term** *recessive* in a complete sentence.

Understand Key Concepts

4. A recessive trait is observed when an organism has _____ recessive genetic factor(s).
   A. 0   C. 2
   B. 1   D. 3

5. **Summarize** Mendel's conclusions about how traits pass from parents to offspring.

6. **Describe** how Mendel cross-pollinated pea plants.

Interpret Graphics

7. **Suppose** the two true-breeding plants shown below were crossed.

What color would the flowers of the offspring be? Explain.

Critical Thinking

8. **Design an experiment** to test for true-breeding plants.

9. **Examine** how Mendel's conclusions disprove blending inheritance.

Math Skills Review — Math Practice —

10. A cross between two pink camellia plants produced the following offspring: 7 plants with red flowers, 7 with white flowers, and 14 with pink flowers. What is the ratio of red to white to pink?

Pioneering the Science of Genetics

SCIENCE & SOCIETY

One man's curiosity leads to a branch of science.

Gregor Mendel—monk, scientist, gardener, and beekeeper—was a keen observer of the world around him. Curious about how traits pass from one generation to the next, he grew and tested almost 30,000 pea plants. Today, Mendel is called the father of genetics. After Mendel published his findings, however, his "laws of heredity" were overlooked for several decades.

In 1900, three European scientists, working independently of one another, rediscovered Mendel's work and replicated his results. Then, other biologists quickly began to recognize the importance of Mendel's work.

▶ Gregor Mendel

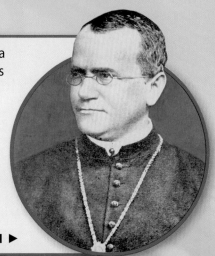

1902: American physician Walter Sutton demonstrates that Mendel's laws of inheritance can be applied to chromosomes. He concludes that chromosomes contain a cell's hereditary material on genes.

1952: American geneticists Martha Chase and Alfred Hershey prove that DNA transmits inherited traits from one generation to the next.

1906: William Bateson, a United Kingdom scientist, coins the term *genetics*. He uses it to describe the study of inheritance and the science of biological inheritance.

1953: Francis Crick and James Watson determine the structure of the DNA molecule. Their work begins the field of molecular biology and leads to important scientific and medical research in genetics.

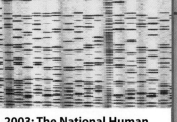

2003: The National Human Genome Research Institute (NHGRI) completes mapping and sequencing human DNA. Researchers and scientists are now trying to discover the genetic basis for human health and disease.

It's Your Turn

RESEARCH What are some genetic diseases? Report on how genome-based research might help cure these diseases in the future.

Lesson 1 EXTEND

Lesson 2

Understanding Inheritance

Reading Guide

Key Concepts
ESSENTIAL QUESTIONS

- What determines the expression of traits?
- How can inheritance be modeled?
- How do some patterns of inheritance differ from Mendel's model?

Vocabulary

gene p. 160
allele p. 160
phenotype p. 160
genotype p. 160
homozygous p. 161
heterozygous p. 161
Punnett square p. 162
incomplete dominance p. 164
codominance p. 164
polygenic inheritance p. 165

Multilingual eGlossary

Video BrainPOP®

Inquiry Make the Connection

Physical traits, such as those shown in these eyes, can vary widely from person to person. Take a closer look at the eyes on this page. What traits can you identify among them? How do they differ?

Inquiry Launch Lab

15 minutes

What is the span of your hand?

Mendel discovered some traits have a simple pattern of inheritance—dominant or recessive. However, some traits, such as eye color, have more variation. Is human hand span a Mendelian trait?

1. Read and complete a lab safety form.
2. Use a **metric ruler** to measure the distance (in cm) between the tips of your thumb and little finger with your hand stretched out.
3. As a class, record everyone's name and hand span in a data table.

Think About This

1. What range of hand span measurements did you observe?
2. **Key Concept** Do you think hand span is a simple Mendelian trait like pea plant flower color?

What controls traits?

Mendel concluded that two factors—one from each parent—control each trait. Mendel hypothesized that one factor came from the egg cell and one factor came from the sperm cell. What are these factors? How are they passed from parents to offspring?

Chromosomes

When other scientists studied the parts of a cell and combined Mendel's work with their work, these factors were more clearly understood. Scientists discovered that inside each cell is a nucleus that contains thread-like structures called chromosomes. Over time, scientists learned that chromosomes contain genetic information that controls traits. We now know that Mendel's "factors" are part of chromosomes and that each cell in offspring contains chromosomes from both parents. As shown in **Figure 5,** these chromosomes exist as pairs—one chromosome from each parent.

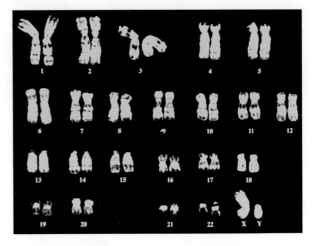

Figure 5 Humans have 23 pairs of chromosomes. Each pair has one chromosome from the father and one chromosome from the mother.

Lesson 2
159
EXPLORE

Genes and Alleles

Scientists have discovered that each chromosome can have information about hundreds or even thousands of traits. *A **gene** (JEEN) is a section on a chromosome that has genetic information for one trait.* For example, a gene of a pea plant might have information about flower color. Recall that an offspring inherits two genes (factors) for each trait—one from each parent. The genes can be the same or different, such as purple or white for pea flower color. *The different forms of a gene are called **alleles** (uh LEELs).* Pea plants can have two purple alleles, two white alleles, or one of each allele. In **Figure 6,** the chromosome pair has information about three traits—flower position, pod shape, and stem length.

 Reading Check How many alleles controlled flower color in Mendel's experiments?

Genotype and Phenotype

Look again at the photo at the beginning of this lesson. What human trait can you observe? You might observe that eye color can be shades of blue or brown. *Geneticists call how a trait appears, or is expressed, the trait's **phenotype** (FEE nuh tipe).* What other phenotypes can you observe in the photo?

Mendel concluded that two alleles control the expression or phenotype of each trait. *The two alleles that control the phenotype of a trait are called the trait's **genotype** (JEE nuh tipe).* Although you cannot see an organism's genotype, you can make inferences about a genotype based on its phenotype. For example, you have already learned that a pea plant with white flowers has two recessive alleles for that trait. These two alleles are its genotype. The white flower is its phenotype.

WORD ORIGIN

phenotype
from Greek *phainein*, means "to show"

Figure 6 The alleles for flower position are the same on both chromosomes. However, the chromosome pair has different alleles for pod shape and stem length.

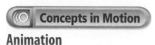
Animation

Chromosome Pair

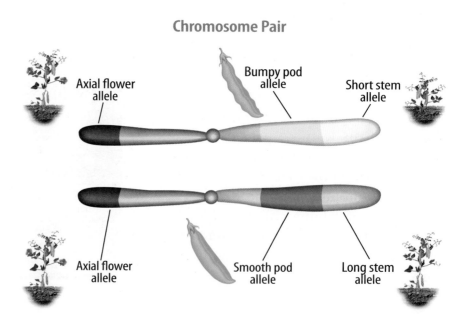

Symbols for Genotypes Scientists use symbols to represent the alleles in a genotype. In genetics, uppercase letters represent dominant alleles and lowercase letters represent recessive alleles. **Table 2** shows the possible genotypes for both round and wrinkled seed phenotypes. Notice that the dominant allele, if present, is written first.

Table 2 Phenotype and Genotype	
Phenotypes (observed traits)	**Genotypes** (alleles of a gene)
Round	Homozygous dominant (RR)
	Heterozygous (Rr)
Wrinkled	Homozygous recessive (rr)

A round seed can have two genotypes—*RR* and *Rr*. Both genotypes have a round phenotype. Why does *Rr* result in round seeds? This is because the round allele *(R)* is dominant to the wrinkled allele *(r)*.

A wrinkled seed has the recessive genotype, *rr*. The wrinkled-seed phenotype is possible only when the same two recessive alleles *(rr)* are present in the genotype.

Homozygous and Heterozygous *When the two alleles of a gene are the same, its genotype is* **homozygous** (hoh muh ZI gus). Both *RR* and *rr* are homozygous genotypes, as shown in **Table 2**.

If the two alleles of a gene are different, its genotype is **heterozygous** (he tuh roh ZI gus). *Rr* is a heterozygous genotype.

Key Concept Check How do alleles determine the expression of traits?

Inquiry MiniLab 20 minutes

Can you infer genotype?

If you know that dragon traits are either dominant or recessive, can you use phenotypes of traits to infer genotypes?

1. Select one **trait card** from each of three **dragon trait bags.** Record the data in your Science Journal.
2. Draw a picture of your dragon based on your data. Label each trait *homozygous* or *heterozygous*.
3. Copy the table below in your Science Journal. For each of the three traits, place one check mark in the appropriate box.

Dragon Traits		
Phenotype	Homozygous	Heterozygous
Green body		
Red body		
Four legs		
Two legs		
Long wings		
Short wings		

4. Combine your data with your classmates' data.

Analyze and Conclude

1. **Describe** any patterns you find in the data table.
2. **Determine** which trait is dominant and which is recessive. Support your reasoning.
3. **Determine** the genotype(s) for each phenotype. Support your reasoning.
4. **Key Concept** Decide whether you could have correctly determined your dragon's genotype without data from other dragons. Support your reasoning.

Punnett Square

Review Personal Tutor

Figure 7 A Punnett square can be used to predict the possible genotypes of the offspring. Offspring from a cross between two heterozygous parents can have one of three genotypes.

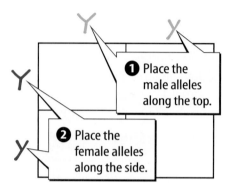

❶ Place the male alleles along the top.

❷ Place the female alleles along the side.

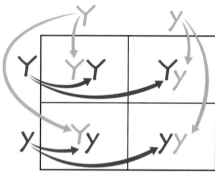

❸ Copy female alleles across each row. Copy male alleles down each column. Always list the dominant trait first.

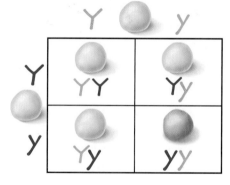

Visual Check What phenotypes are possible for pea offspring of this cross?

Modeling Inheritance

Have you ever flipped a coin and guessed heads or tails? Because a coin has two sides, there are only two possible outcomes—heads or tails. You have a 50 percent chance of getting heads and a 50 percent chance of getting tails. The chance of getting an outcome can be represented by a ratio. The ratio of heads to tails is 50:50 or 1:1.

 Reading Check What does a ratio of 2:1 mean?

Plant breeders and animal breeders use a method for predicting how often traits will appear in offspring that does not require performing the crosses thousands of times. Two models—a Punnett square and a pedigree—can be used to predict and identify traits among genetically related individuals.

Punnett Squares

If the genotypes of the parents are known, then the different genotypes and phenotypes of the offspring can be predicted. *A* **Punnett square** *is a model used to predict possible genotypes and phenotypes of offspring.* Follow the steps in **Figure 7** to learn how to make a Punnett square.

Analyzing a Punnett Square

Figure 7 shows an example of a cross between two pea plants that are heterozygous for pea seed color—*Yy* and *Yy*. Yellow is the dominant allele—*Y*. Green is the recessive allele—*y*. The offspring can have one of three genotypes—*YY*, *Yy*, or *yy*. The ratio of genotypes is written as 1:2:1.

Because *YY* and *Yy* represent the same phenotype—yellow—the offspring can have one of only two phenotypes—yellow or green. The ratio of phenotypes is written 3:1. Therefore, about 75 percent of the offspring of the cross between two heterozygous pea plants will produce yellow seeds. About 25 percent of the plants will produce green seeds.

Using Ratios to Predict

Given a 3:1 ratio, you can expect that an offspring from heterozygous parents has a 3:1 chance of having yellow seeds. But you cannot expect that a group of four seeds will have three yellow seeds and one green seed. This is because one offspring does not affect the phenotype of another offspring. In a similar way, the outcome of one coin toss does not affect the outcome of other coin tosses.

However, if you counted large numbers of offspring from a particular cross, the overall ratio would be close to the ratio predicted by a Punnett square. Mendel did not use Punnett squares. However, by studying nearly 30,000 pea plants, his ratios nearly matched those that would have been predicted by a Punnett square for each cross.

Pedigrees

Another model that can show inherited traits is a pedigree. A pedigree shows phenotypes of genetically related family members. It can also help determine genotypes. In the pedigree in **Figure 8,** three offspring have a trait—attached earlobes—that the parents do not have. If these offspring received one allele for this trait from each parent, but neither parent displays the trait, the offspring must have received two recessive alleles.

 Key Concept Check How can inheritance be modeled?

Pedigree

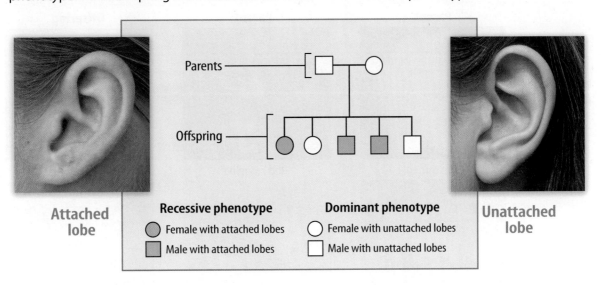

Figure 8 In this pedigree, the parents and two offspring have unattached ear lobes—the dominant phenotype. Three offspring have attached ear lobes—the recessive phenotype.

Visual Check If the genotype of the offspring with attached lobes is *uu,* what is the genotype of the parents? How can you tell?

Complex Patterns of Inheritance

By chance, Mendel studied traits only influenced by one gene with two alleles. However, we know now that some inherited traits have complex patterns of inheritance.

Types of Dominance

Recall that for pea plants, the presence of one dominant allele produces a dominant phenotype. However, not all allele pairs have a dominant-recessive interaction.

Incomplete Dominance Sometimes traits appear to be combinations of alleles. *Alleles show* **incomplete dominance** *when the offspring's phenotype is a combination of the parents' phenotypes.* For example, a pink camellia, as shown in **Figure 9,** results from incomplete dominance. A cross between a camellia plant with white flowers and a camellia plant with red flowers produces only camellia plants with pink flowers.

Codominance The coat color of some cows is an example of another type of interaction between two alleles. *When both alleles can be observed in a phenotype, this type of interaction is called* **codominance.** If a cow inherits the allele for white coat color from one parent and the allele for red coat color from the other parent, the cow will have both red and white hairs.

FOLDABLES

Use two sheets of paper to make a layered book. Label it as shown. Use it to organize your notes on inheritance patterns.

- Inheritance Patterns
- Incomplete dominance
- Multiple alleles
- Polygenic inheritance

Types of Dominance

Figure 9 In incomplete dominance, neither parent's phenotype is visible in the offspring's phenotype. In codominance, both parents' phenotypes are visible separately in the offspring's phenotype.

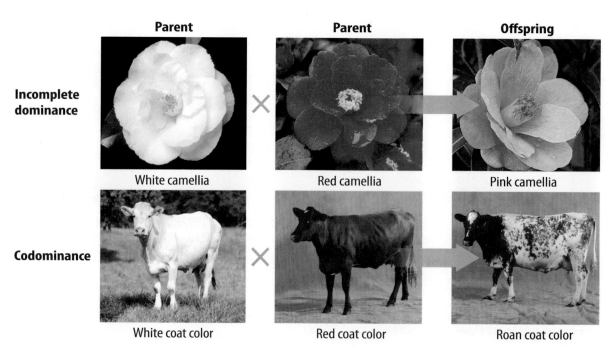

Table 3 Human ABO Blood Types

Phenotype	Possible Genotypes
Type A	$I^A I^A$ or $I^A i$
Type B	$I^B I^B$ or $I^B i$
Type O	ii
Type AB	$I^A I^B$

Multiple Alleles

Unlike the genes in Mendel's pea plants, some genes have more than two alleles, or multiple alleles. Human ABO blood type is an example of a trait that is determined by multiple alleles. There are three different alleles for the ABO blood type—I^A, I^B, and i. The way the alleles combine results in one of four blood types—A, B, AB, or O. The I^A and I^B alleles are codominant to each other, but they both are dominant to the i allele. Even though there are multiple alleles, a person can inherit only two of these alleles—one from each parent, as shown in **Table 3**.

Polygenic Inheritance

Mendel concluded that each trait was determined by only one gene. However, we now know that a trait can be affected by more than one gene. **Polygenic inheritance** *occurs when multiple genes determine the phenotype of a trait.* Because several genes determine a trait, many alleles affect the phenotype even though each gene has only two alleles. Therefore, polygenic inheritance has many possible phenotypes.

Look again at the photo at the beginning of this lesson. Eye color in humans is an example of polygenic inheritance. There are also many phenotypes for height in humans, as shown in **Figure 10**. Other human characteristics determined by polygenic inheritance are weight and skin color.

ACADEMIC VOCABULARY
conclude
(verb) to reach a logically necessary end by reasoning

 Key Concept Check How does polygenic inheritance differ from Mendel's model?

Figure 10 The eighth graders in this class have different heights.

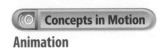
Animation

Genes and the Environment

You read earlier in this lesson that an organism's genotype determines its phenotype. Scientists have learned that genes are not the only factors that can affect phenotypes. An organism's environment can also affect its phenotype. For example, the flower color of one type of hydrangea is determined by the soil in which the hydrangea plant grows. **Figure 11** shows that acidic soil produces blue flowers and basic, or alkaline, soil produces pink flowers. Other examples of environmental effects on phenotype are also shown in **Figure 11**.

For humans, healthful choices can also affect phenotype. Many genes affect a person's chances of having heart disease. However, what a person eats and the amount of exercise he or she gets can influence whether heart disease will develop.

Reading Check What environmental factors affect phenotype?

Figure 11 Environmental factors, such as temperature and sunlight, can affect phenotype.

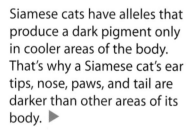

◀ These hydrangea plants are genetically identical. The plant grown in acidic soil produced blue flowers. The plant grown in alkaline soil produced pink flowers.

Siamese cats have alleles that produce a dark pigment only in cooler areas of the body. That's why a Siamese cat's ear tips, nose, paws, and tail are darker than other areas of its body. ▶

◀ The wing patterns of the map butterfly, *Araschnia levana*, depend on what time of year the adult develops. Adults that developed in the spring have more orange in their wings than those that developed in the summer.

Lesson 2 Review

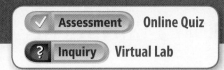

Visual Summary

The genes for traits are located on chromosomes.

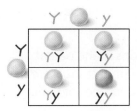

Geneticists use Punnett squares to predict the possible genotypes and phenotypes of offspring.

In polygenic inheritance, traits are determined by more than one gene and have many possible phenotypes.

FOLDABLES

Use your lesson Foldable to review the lesson. Save your Foldable for the project at the end of the chapter.

What do you think NOW?

You first read the statements below at the beginning of the chapter.

3. All inherited traits follow Mendel's patterns of inheritance.

4. Scientists have tools to predict the form of a trait an offspring might inherit.

Did you change your mind about whether you agree or disagree with the statements? Rewrite any false statements to make them true.

Use Vocabulary

① **Use** the terms *phenotype* and *genotype* in a complete sentence.

② **Contrast** homozygous and heterozygous.

③ **Define** *incomplete dominance* in your own words.

Understand Key Concepts

④ How many alleles control a Mendelian trait, such as pea seed color?
   A. one   C. three
   B. two   D. four

⑤ **Explain** where the alleles for a given trait are inherited from.

⑥ **Describe** how the genotypes *RR* and *Rr* result in the same phenotype.

⑦ **Summarize** how polygenic inheritance differs from Mendelian inheritance.

Interpret Graphics

⑧ **Analyze** this pedigree. If ■ represents a male with the homozygous recessive genotype (*aa*), what is the mother's genotype?

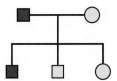

Critical Thinking

⑨ **Predict** the possible blood genotypes of a child, using the table below, if one parent is type O and the other parent is type A.

Phenotype	Genotype
Blood Type O	ii
Blood Type A	$I^A I^A$ or $I^A i$

Lesson 2 • **167**
EVALUATE

Inquiry Skill Practice — Model

25 minutes

How can you use Punnett squares to model inheritance?

Geneticists use models to explain how traits are inherited from one generation to the next. A simple model of Mendelian inheritance is a Punnett square. A Punnett square is a model of reproduction between two parents and the possible genotypes and phenotypes of the resulting offspring. It also models the probability that each genotype will occur.

Learn It

In science, a **model** is a representation of how something in the natural world works. A model is used to explain or predict a natural process. Maps, diagrams, three-dimensional representations, and mathematical formulas can all be used to help model nature.

Try It

1. Copy the Punnett square on this page in your Science Journal. Use it to complete a cross between a fruit fly with straight wings (cc) and a fruit fly with curly wings (CC).

2. According to your Punnett square, which genotypes are possible in the offspring?

3. Using the information in your Punnett square, calculate the ratio of the dominant phenotype to the recessive phenotype in the offspring.

Apply It

4. Based on the information in your Punnett square, how many offspring will have curly wings? Straight wings?

5. If you switch the locations of the parent genotypes around the Punnett square, does it affect the potential genotypes of their offspring? Explain.

6. **Key Concept** Design and complete a Punnett square to model a cross between two fruit flies that are heterozygous for the curly wings (Cc). What are the phenotypic ratios of the offspring?

Magnification: 20×

Curly wings (CC)

Straight wings (cc)

168 Chapter 5 EXTEND

Lesson 3

DNA and Genetics

Reading Guide

Key Concepts 🔑
ESSENTIAL QUESTIONS

- What is DNA?
- What is the role of RNA in protein production?
- How do changes in the sequence of DNA affect traits?

Vocabulary

DNA p. 170
nucleotide p. 171
replication p. 172
RNA p. 173
transcription p. 173
translation p. 174
mutation p. 175

g Multilingual eGlossary

Inquiry   What are these coils?

What color are your eyes? How tall are you? Traits are controlled by genes. But genes never leave the nucleus of the cell. How does a gene control a trait? These stringy coils hold the answer to that question.

Inquiry Launch Lab

20 minutes

How are codes used to determine traits?

Interpret this code to learn more about how an organism's body cells use codes to determine genetic traits.

1. Analyze the pattern of the simple code shown to the right. For example, ⟩⟨⌐ = DOG

2. In your Science Journal, record the correct letters for the symbols in the code below.

Think About This

1. What do all codes, such as Morse code and Braille, have in common?
2. What do you think might happen if there is a mistake in the code?
3. **Key Concept** How do you think an organism's cells might use code to determine its traits?

The Structure of DNA

Have you ever put together a toy or a game for a child? If so, it probably came with directions. Cells put molecules together in much the same way you might assemble a toy. They follow a set of directions.

Genes provide directions for a cell to assemble molecules that express traits such as eye color or seed shape. Recall from Lesson 2 that a gene is a section of a chromosome. Chromosomes are made of proteins and deoxyribonucleic (dee AHK sih ri boh noo klee ihk) acid, or **DNA**—*an organism's genetic material*. A gene is a segment of DNA on a chromosome.

Cells and organisms contain millions of different molecules. Countless numbers of directions are needed to make all those molecules. How do all these directions fit on a few chromosomes? The information, or directions, needed for an organism to grow, maintain itself, and reproduce is contained in DNA. As shown in **Figure 12**, strands of DNA in a chromosome are tightly coiled, like a telephone cord or a coiled spring. This coiling allows more genes to fit in a small space.

Key Concept Check What is DNA?

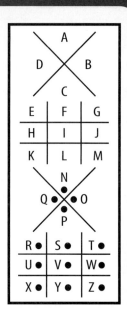

Figure 12 Strands of DNA are tightly coiled in chromosomes.

A Complex Molecule

What's the best way to fold clothes so they will fit into a drawer or a suitcase? Scientists asked a similar question about DNA. What is the shape of the DNA molecule, and how does it fit into a chromosome? The work of several scientists revealed that DNA is like a twisted zipper. This twisted zipper shape is called a double helix. A model of DNA's double helix structure is shown in **Figure 13**.

How did scientists make this discovery? Rosalind Franklin and Maurice Wilkins were two scientists in London who used X-rays to study DNA. Some of the X-ray data indicated that DNA has a helix shape.

American scientist James Watson visited Franklin and Wilkins and saw one of the DNA X-rays. Watson realized that the X-ray gave valuable clues about DNA's structure. Watson worked with an English scientist, Francis Crick, to build a model of DNA.

Watson and Crick based their work on information from Franklin's and Wilkins's X-rays. They also used chemical information about DNA discovered by another scientist, Erwin Chargaff. After several tries, Watson and Crick built a model that showed how the smaller molecules of DNA bond together and form a double helix.

Four Nucleotides Shape DNA

DNA's twisted-zipper shape is because of molecules called nucleotides. *A **nucleotide** is a molecule made of a nitrogen base, a sugar, and a phosphate group.* Sugar-phosphate groups form the sides of the DNA zipper. The nitrogen bases bond and form the teeth of the zipper. As shown in **Figure 13**, there are four nitrogen bases: adenine (A), cytosine (C), thymine (T), and guanine (G). A and T always bond together, and C and G always bond together.

✓ **Reading Check** What is a nucleotide?

Figure 13 A DNA double helix is made of two strands of DNA. Each strand is a chain of nucleotides.

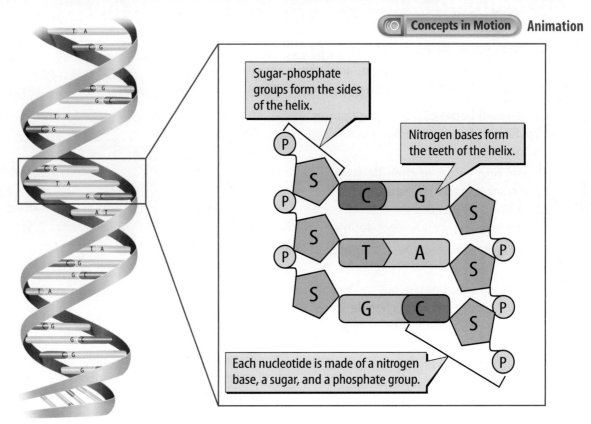

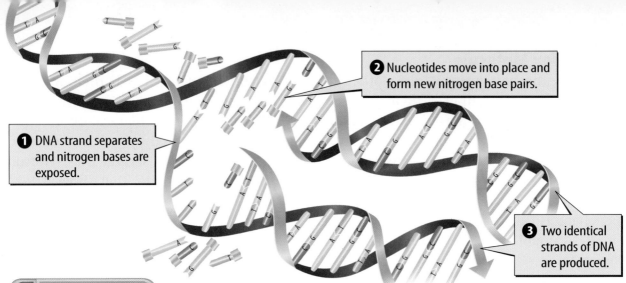

① DNA strand separates and nitrogen bases are exposed.

② Nucleotides move into place and form new nitrogen base pairs.

③ Two identical strands of DNA are produced.

Concepts in Motion
Animation

Figure 14 Before a cell divides, its DNA is replicated.

How DNA Replicates

Cells contain DNA in chromosomes. So, every time a cell divides, all chromosomes must be copied for the new cell. The new DNA is identical to existing DNA. *The process of copying a DNA molecule to make another DNA molecule is called* **replication**. You can follow the steps of DNA replication in **Figure 14.** First, the strands separate in many places, exposing individual bases. Then nucleotides are added to each exposed base. This produces two identical strands of DNA.

Reading Check What is replication?

Inquiry MiniLab
25 minutes

How can you model DNA?

Making a model of DNA can help you understand its structure.

1. Read and complete a lab safety form.
2. Link a **small paper clip** to a **large paper clip.** Repeat four more times, making a chain of 10 paper clips.
3. Choose **four colors of chenille stems.** Each color represents one of the four nitrogen bases. Record the color of each nitrogen base in your Science Journal.
4. Attach a chenille stem to each large paper clip.
5. Repeat step 2 and step 4, but this time attach the corresponding chenille-stem nitrogen bases. Connect the nitrogen bases.
6. Securely insert one end of your double chain into a **block of styrene foam.**
7. Repeat step 6 with the other end of your chain.
8. Gently turn the blocks to form a double helix.

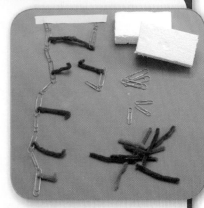

Analyze and Conclude

1. **Explain** which part of a DNA molecule is represented by each material you used.
2. **Predict** what might happen if a mistake were made in creating a nucleotide.
3. **Key Concept** How did making a model of DNA help you understand its structure?

Making Proteins

Recall that proteins are important for every cellular process. The DNA of each cell carries a complete set of genes that provides instructions for making all the proteins a cell requires. Most genes contain instructions for making proteins. Some genes contain instructions for when and how quickly proteins are made.

Junk DNA

As you have learned, all genes are segments of DNA on a chromosome. However, you might be surprised to learn that most of your DNA is not part of any gene. For example, about 97 percent of the DNA on human chromosomes does not form genes. Segments of DNA that are not parts of genes are often called junk DNA. It is not yet known whether junk DNA segments have functions that are important to cells.

The Role of RNA in Making Proteins

How does a cell use the instructions in a gene to make proteins? Proteins are made with the help of ribonucleic acid (**RNA**)—*a type of nucleic acid that carries the code for making proteins from the nucleus to the cytoplasm.* RNA also carries amino acids around inside a cell and forms a part of ribosomes.

RNA, like DNA, is made of nucleotides. However, there are key differences between DNA and RNA. DNA is double-stranded, but RNA is single-stranded. RNA has the nitrogen base uracil (U) instead of thymine (T) and the sugar ribose instead of deoxyribose.

The first step in making a protein is to make mRNA from DNA. *The process of making mRNA from DNA is called* **transcription.** **Figure 15** shows how mRNA is transcribed from DNA.

 Key Concept Check What is the role of RNA in protein production?

Transcription

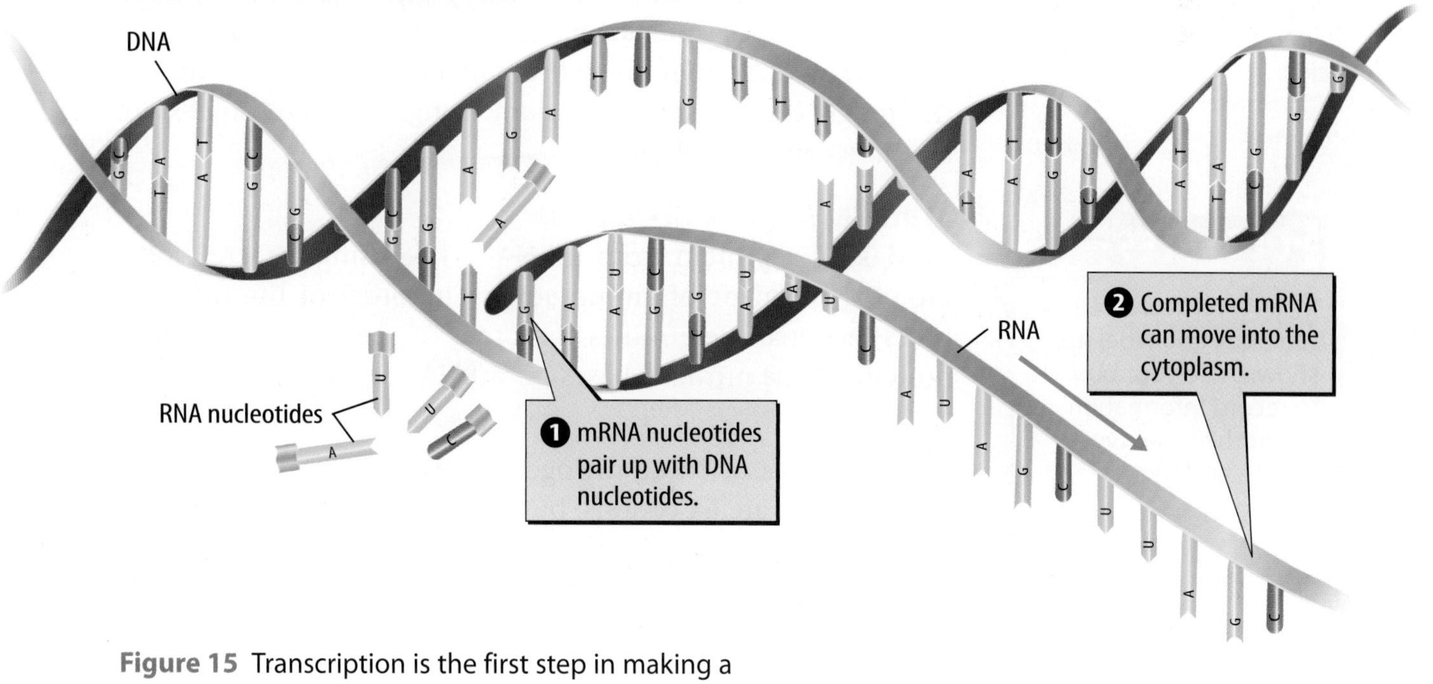

Figure 15 Transcription is the first step in making a protein. During transcription, the sequence of nitrogen bases on a gene determines the sequence of bases on mRNA.

Lesson 3
EXPLAIN
173

Translation

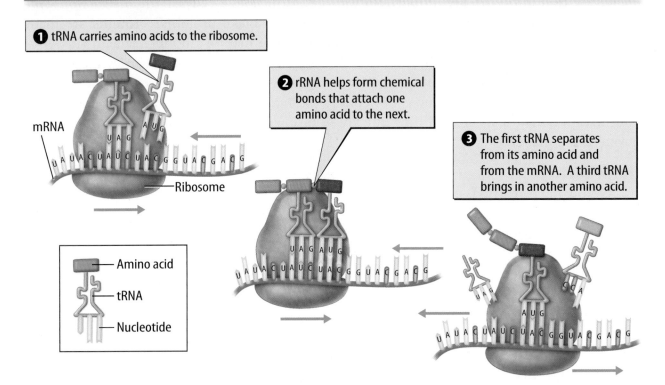

Figure 16 A protein forms as mRNA moves through a ribosome. Different amino acid sequences make different proteins. A complete protein is a folded chain of amino acids.

FOLDABLES

Make a vertical three-tab book and label it as shown. Use your book to record information about the three types of RNA and their functions.

Three Types of RNA

On the previous page, you read about messenger RNA (mRNA). There are two other types of RNA, transfer RNA (tRNA) and ribosomal RNA (rRNA). **Figure 16** illustrates how the three work together to make proteins. *The process of making a protein from RNA is called* **translation.** Translation occurs in ribosomes. Recall that ribosomes are cell organelles that are attached to the rough endoplasmic reticulum (rough ER). Ribosomes are also in a cell's cytoplasm.

Translating the RNA Code

Making a protein from mRNA is like using a secret code. Proteins are made of amino acids. The order of the nitrogen bases in mRNA determines the order of the amino acids in a protein. Three nitrogen bases on mRNA form the code for one amino acid.

Each series of three nitrogen bases on mRNA is called a codon. There are 64 codons, but only 20 amino acids. Some of the codons code for the same amino acid. One of the codons codes for an amino acid that is the beginning of a protein. This codon signals that translation should start. Three of the codons do not code for any amino acid. Instead, they code for the end of the protein. They signal that translation should stop.

 Reading Check What is a codon?

174 • Chapter 5
EXPLAIN

Mutations

You have read that the sequence of nitrogen bases in DNA determines the sequence of nitrogen bases in mRNA, and that the mRNA sequence determines the sequence of amino acids in a protein. You might think these sequences always stay the same, but they can change. *A change in the nucleotide sequence of a gene is called a* **mutation.**

The 46 human chromosomes contain between 20,000 and 25,000 genes that are copied during DNA replication. Sometimes, mistakes can happen during replication. Most mistakes are corrected before replication is completed. A mistake that is not corrected can result in a mutation. Mutations can be triggered by exposure to X-rays, ultraviolet light, radioactive materials, and some kinds of chemicals.

Types of Mutations

There are several types of DNA mutations. Three types are shown in **Figure 17.** In a deletion mutation, one or more nitrogen bases are left out of the DNA sequence. In an insertion mutation, one or more nitrogen bases are added to the DNA. In a substitution mutation, one nitrogen base is replaced by a different nitrogen base.

Each type of mutation changes the sequence of nitrogen base pairs. This can cause a mutated gene to code for a different protein than a normal gene. Some mutated genes do not code for any protein. For example, a cell might lose the ability to make one of the proteins it needs.

WORD ORIGIN
mutation
from Latin *mutare*, means "to change"

Figure 17 Three types of mutations are substitution, insertion, and deletion.

Visual Check Which base pairs were omitted during replication in the deletion mutation?

Mutations

Original DNA sequence

Substitution
The C-G base pair has been replaced with a T-A pair.

Insertion
Three base pairs have been added.

Deletion
Three base pairs have been removed. Other base pairs will move in to take their place.

Results of a Mutation

The effects of a mutation depend on where in the DNA sequence the mutation happens and the type of mutation. Proteins express traits. Because mutations can change proteins, they can cause traits to change. Some mutations in human DNA cause genetic disorders, such as those described in **Table 4.**

However, not all mutations have negative effects. Some mutations don't cause changes in proteins, so they don't affect traits. Other mutations might cause a trait to change in a way that benefits the organism.

 Key Concept Check How do changes in the sequence of DNA affect traits?

Scientists still have much to learn about genes and how they determine an organism's traits. Scientists are researching and experimenting to identify all genes that cause specific traits. With this knowledge, we might be one step closer to finding cures and treatments for genetic disorders.

Table 4 Genetic Disorders

Defective Gene or Chromosome	Disorder	Description
Chromosome 12, PAH gene	Phenylketonuria (PKU)	People with defective PAH genes cannot break down the amino acid phenylalanine. If phenylalanine builds up in the blood, it poisons nerve cells.
Chromosome 7, CFTR gene	Cystic fibrosis	In people with defective CFTR genes, salt cannot move in and out of cells normally. Mucus builds up outside cells. The mucus can block airways in lungs and affect digestion.
Chromosome 7, elastin gene	Williams syndrome	People with Williams syndrome are missing part of chromosome 7, including the elastin gene. The protein made from the elastin gene makes blood vessels strong and stretchy.
Chromosome 17, BRCA 1; Chromosome 13, BRCA 2	Breast cancer and ovarian cancer	A defect in BRCA1 and/or BRCA2 does not mean the person will have breast cancer or ovarian cancer. People with defective BRCA1 or BRCA2 genes have an increased risk of developing breast cancer and ovarian cancer.

Lesson 3 Review

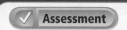

Visual Summary

DNA is a complex molecule that contains the code for an organism's genetic information.

RNA carries the codes for making proteins.

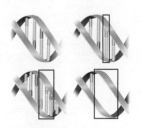

An organism's nucleotide sequence can change through the deletion, insertion, or substitution of nitrogen bases.

FOLDABLES

Use your lesson Foldable to review the lesson. Save your Foldable for the project at the end of the chapter.

What do you think NOW?

You first read the statements below at the beginning of the chapter.

5. Any condition present at birth is genetic.

6. A change in the sequence of an organism's DNA always changes the organism's traits.

Did you change your mind about whether you agree or disagree with the statements? Rewrite any false statements to make them true.

Use Vocabulary

1 **Distinguish** between transcription and translation.

2 **Use the terms** *DNA* and *nucleotide* in a sentence.

3 A change in the sequence of nitrogen bases in a gene is called a(n) _____.

Understand Key Concepts

4 Where does the process of transcription occur?
   A. cytoplasm   C. cell nucleus
   B. ribosomes   D. outside the cell

5 **Illustrate** Make a drawing that illustrates the process of translation.

6 **Distinguish** between the sides of the DNA double helix and the teeth of the DNA double helix.

Interpret Graphics

7 **Identify** The products of what process are shown in the figure below?

8 **Sequence** Draw a graphic organizer like the one below about important steps in making a protein, beginning with DNA and ending with protein.

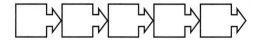

Critical Thinking

9 **Hypothesize** What would happen if a cell were unable to make mRNA?

10 **Assess** What is the importance of DNA replication occurring without any mistakes?

Lesson 3 EVALUATE

Inquiry Lab

40 minutes

Gummy Bear Genetics

Materials

gummy bears

calculator

paper bag

Safety

Imagine you are on a team of geneticists that is doing "cross-breeding experiments" with gummy bears. Unfortunately, the computer containing your data has crashed. All you have left are six gummy-bear litters that resulted from six sets of parents. But no one can remember which parents produced which litter. You know that gummy-bear traits have either Mendelian inheritance or incomplete dominance. Can you determine which parents produced each set of offspring and how gummy bear traits are inherited?

Ask a Question

What are the genotypes and phenotypes of the parents for each litter?

Make Observations

1. Obtain a bag of gummy bears. Sort the bears by color (phenotype).

   ⚠ *Do not eat the gummy bears.*

2. Count the number (frequency) of bears for each phenotype. Then, calculate the ratio of phenotypes for each litter.

3. Combine data from your litter with those of your classmates using a data table like the one below.

4. As a class, select a letter to represent the alleles for color. Record the possible genotypes for your bears in the class data table.

Gummy Bear Cross Data for Lab Group

Cross #	Phenotype Frequencies	Ratio	Possible Genotypes	Mode of Inheritance	Predicted Parental Genotypes
EXAMPLE	15 green/5 pink	3:1	GG or Gg/gg	Mendelian	Gg x Gg
1.					
2.					
3.					
4.					

Form a Hypothesis

5. Use the data to form a hypothesis about the probable genotypes and phenotypes of the parents of your litter and the probable type of inheritance.

Test Your Hypothesis

6. Design and complete a Punnett square using the predicted parental genotypes in your hypothesis.

7. Compare your litter's phenotype ratio with the ratio predicted by the Punnett square. Do your data support your hypothesis? If not, revise your hypothesis and repeat steps 5–7.

Analyze and Conclude

8. **Infer** What were the genotypes of the parents? The phenotypes? How do you know?

9. **The Big Idea** Determine the probable modes of inheritance for each phenotype. Explain your reasoning.

10. **Graph** Using the data you collected, draw a bar graph that compares the phenotype frequency for each gummy bear phenotype.

Reminder
Using Ratios
- ☑ A ratio is a comparison of two numbers.
- ☑ A ratio of 15 : 5 can be reduced to 3 : 1.

Communicate Your Results

Create a video presentation of the results of your lab. Describe the question you investigated, the steps you took to answer the question, and the data that support your conclusions. Share your video with your classmates.

Think of a question you have about genetics. For example, can you design a pedigree to trace a Mendelian trait in your family? To investigate your question, design a controlled experiment or an observational study.

Remember to use scientific methods.

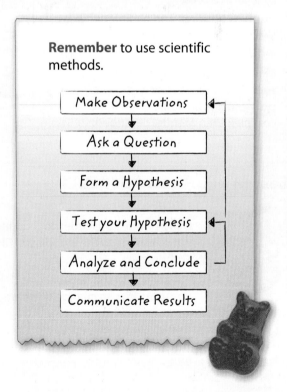

Chapter 5 Study Guide

 Inherited genes are the basis of an organism's traits.

Key Concepts Summary

Lesson 1: Mendel and His Peas
- Mendel performed cross-pollination experiments to track which traits were produced by specific parental crosses.
- Mendel found that two genetic factors—one from a sperm cell and one from an egg cell—control each trait.
- **Dominant** traits block the expression of **recessive** traits. Recessive traits are expressed only when two recessive factors are present.

Lesson 2: Understanding Inheritance
- **Phenotype** describes how a trait appears.
- **Genotype** describes alleles that control a trait.
- **Punnett squares** and pedigrees are tools to model patterns of inheritance.
- Many patterns of inheritance, such as **codominance** and **polygenic inheritance**, are more complex than Mendel described.

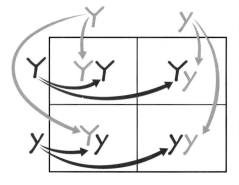

Lesson 3: DNA and Genetics
- **DNA** contains an organism's genetic information.
- **RNA** carries the codes for making proteins from the nucleus to the cytoplasm. RNA also forms part of ribosomes.
- A change in the sequence of DNA, called a **mutation,** can change the traits of an organism.

Vocabulary

heredity p. 149
genetics p. 149
dominant trait p. 155
recessive trait p. 155

gene p. 160
allele p. 160
phenotype p. 160
genotype p. 160
homozygous p. 161
heterozygous p. 161
Punnett square p. 162
incomplete dominance p. 164
codominance p. 164
polygenic inheritance p. 165

DNA p. 170
nucleotide p. 171
replication p. 172
RNA p. 173
transcription p. 173
translation p. 174
mutation p. 175

Study Guide

- Personal Tutor
- Vocabulary eGames
- Vocabulary eFlashcards

Chapter Project

Assemble your lesson Foldables as shown to make a Chapter Project. Use the project to review what you have learned in this chapter.

Use Vocabulary

1. The study of how traits are passed from parents to offspring is called _____.

2. The passing of traits from parents to offspring is _____.

3. Human height, weight, and skin color are examples of characteristics determined by _____ _____.

4. A helpful device for predicting the ratios of possible genotypes is a(n) _____.

5. The code for a protein is called a(n) _____.

6. An error made during the copying of DNA is called a(n) _____.

Link Vocabulary and Key Concepts

 Interactive Concept Map

Copy this concept map, and then use vocabulary terms from the previous page to complete the concept map.

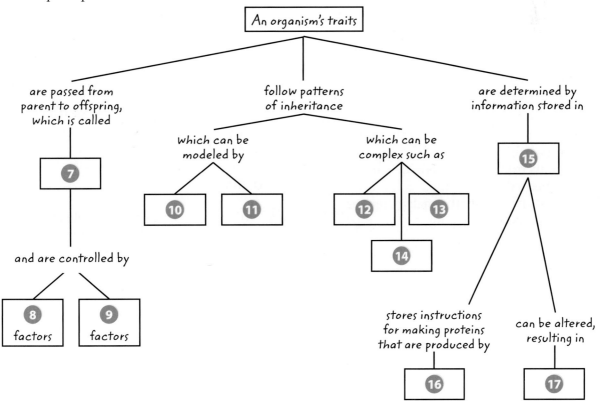

Chapter 5 Review

Understand Key Concepts

1 The process shown below was used by Mendel during his experiments.

What is the process called?
A. cross-pollination
B. segregation
C. asexual reproduction
D. blending inheritance

2 Which statement best describes Mendel's experiments?
A. He began with hybrid plants.
B. He controlled pollination.
C. He observed only one generation.
D. He used plants that reproduce slowly.

3 Before Mendel's discoveries, which statement describes how people believed traits were inherited?
A. Parental traits blend like colors of paint to produce offspring.
B. Parental traits have no effect on their offspring.
C. Traits from only the female parent are inherited by offspring.
D. Traits from only the male parent are inherited by offspring.

4 Which term describes the offspring of a first-generation cross between parents with different forms of a trait?
A. genotype
B. hybrid
C. phenotype
D. true-breeding

5 Which process makes a copy of a DNA molecule?
A. mutation
B. replication
C. transcription
D. translation

6 Which process uses the code on an RNA molecule to make a protein?
A. mutation
B. replication
C. transcription
D. translation

7 The Punnett square below shows a cross between a pea plant with yellow seeds and a pea plant with green seeds.

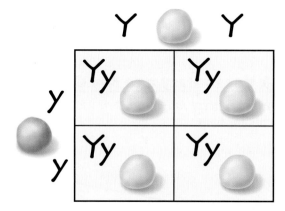

If mating produces 100 offspring, about how many will have yellow seeds?
A. 25
B. 50
C. 75
D. 100

8 Which term describes multiple genes affecting the phenotype of one trait?
A. codominance
B. blending inheritance
C. incomplete dominance
D. polygenic inheritance

Chapter Review

Assessment — Online Test Practice

Critical Thinking

9. Compare heterozygous genotype and homozygous genotype.

10. Distinguish between multiple alleles and polygenic inheritance.

11. Give an example of how the environment can affect an organism's phenotype.

12. Predict In pea plants, the allele for smooth pods is dominant to the allele for bumpy pods. Predict the genotype of a plant with bumpy pods. Can you predict the genotype of a plant with smooth pods? Explain.

13. Interpret Graphics In tomato plants, red fruit (R) is dominant to yellow fruit (r). Interpret the Punnett square below, which shows a cross between a heterozygous red plant and a yellow plant. Include the possible genotypes and corresponding phenotypes.

	R	r
r	Rr	rr
r	Rr	rr

14. Compare and contrast characteristics of replication, transcription, translation, and mutation. Which of these processes takes place only in the nucleus of a cell? Which can take place in both the nucleus and the cytoplasm? How do you know?

Writing in Science

15. Write a paragraph contrasting the blending theory of inheritance with the current theory of inheritance. Include a main idea, supporting details, and a concluding sentence.

REVIEW THE BIG IDEA

16. How are traits passed from generation to generation? Explain how dominant and recessive alleles interact to determine the expression of traits.

17. The photo below shows an albino offspring from a non-albino mother. If albinism is a recessive trait, what are the possible genotypes of the mother, the father, and the offspring?

Math Skills

Review — Math Practice

Use Ratios

18. A cross between two heterozygous pea plants with yellow seeds produced 1,719 yellow seeds and 573 green seeds. What is the ratio of yellow to green seeds?

19. A cross between two heterozygous pea plants with smooth green pea pods produced 87 bumpy yellow pea pods, 261 smooth yellow pea pods, 261 bumpy green pea pods, and 783 smooth green pea pods. What is the ratio of bumpy yellow to smooth yellow to bumpy green to smooth green pea pods?

20. A jar contains three red, five green, two blue, and six yellow marbles. What is the ratio of red to green to blue to yellow marbles?

Standardized Test Practice

Record your answers on the answer sheet provided by your teacher or on a sheet of paper.

Multiple Choice

Use the diagram below to answer questions 1 and 2.

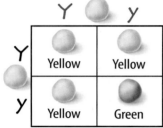

1 Which genotype belongs in the lower right square?

   A YY
   B Yy
   C yY
   D yy

2 What percentage of plants from this cross will produce yellow seeds?

   A 25 percent
   B 50 percent
   C 75 percent
   D 100 percent

3 When Mendel crossed a true-breeding plant with purple flowers and a true-breeding plant with white flowers, ALL offspring had purple flowers. This is because white flowers are

   A dominant.
   B heterozygous.
   C polygenic.
   D recessive.

4 Which process copies an organism's DNA?

   A mutation
   B replication
   C transcription
   D translation

Use the chart below to answer question 5.

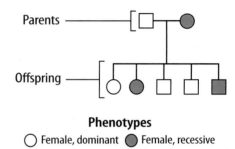

5 Based on the pedigree above, how many offspring from this cross had the recessive phenotype?

   A 1
   B 2
   C 3
   D 5

6 Which is NOT true of a hybrid?

   A It has one recessive allele.
   B It has pairs of chromosomes.
   C Its genotype is homozygous.
   D Its phenotype is dominant.

7 Alleles are different forms of a

   A chromosome.
   B gene.
   C nucleotide.
   D protein.

8 Which is true of an offspring with incomplete dominance?

   A Both alleles can be observed in its phenotype.
   B Every offspring shows the dominant phenotype.
   C Multiple genes determine its phenotype.
   D Offspring phenotype is a combination of the parents' phenotypes.

Standardized Test Practice

Use the diagrams below to answer question 9.

Before Replication

After Replication

9. The diagrams above show a segment of DNA before and after replication. Which occurred during replication?

   A deletion

   B insertion

   C substitution

   D translation

10. Which human characteristic is controlled by polygenic inheritance?

    A blood type

    B earlobe position

    C eye color

    D thumb shape

11. Mendel crossed a true-breeding plant with round seeds and a true-breeding plant with wrinkled seeds. Which was true of every offspring of this cross?

    A They had the recessive phenotype.

    B They showed a combination of traits.

    C They were homozygous.

    D They were hybrid plants.

Constructed Response

Use the diagram below to answer questions 12 and 13.

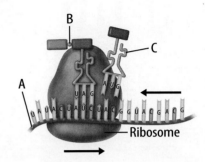

12. Describe what is happening in the phase of translation shown in the diagram.

13. What are the three types of RNA in the diagram? How do these types work together during translation?

14. What is the importance of translation in your body?

15. Mendel began his experiments with true-breeding plants. Why was this important?

16. How did Mendel's experimental methods help him develop his hypotheses on inheritance?

17. What environmental factors affect the phenotypes of organisms other than humans? Provide three examples from nature. What factor, other than genes, affects human phenotype? Give two examples. Why is knowledge of this non-genetic factor helpful?

NEED EXTRA HELP?																	
If You Missed Question...	1	2	3	4	5	6	7	8	9	10	11	12	13	14	15	16	17
Go to Lesson...	2	2	1	3	2	1,2	2	2	3	2	1	3	3	3	1	1	2

Chapter 6
The Environment and Change Over Time

 How do species adapt to changing environments over time?

Inquiry Swarm of Bees?

A type of orchid plant, called a bee orchid, produces this flower. You might have noticed that the flower looks like a bee.

- What is the advantage to the plant to have flowers that look like bees?
- How did the appearance of the flower develop over time?
- How do species adapt to changing environments over time?

Get Ready to Read

What do you think?

Before you read, decide if you agree or disagree with each of these statements. As you read this chapter, see if you change your mind about any of the statements.

1. Original tissues can be preserved as fossils.
2. Organisms become extinct only in mass extinction events.
3. Environmental change causes variations in populations.
4. Variations can lead to adaptations.
5. Living species contain no evidence that they are related to each other.
6. Plants and animals share similar genes.

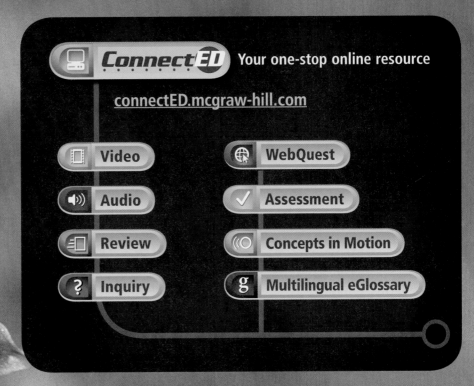

ConnectED Your one-stop online resource

connectED.mcgraw-hill.com

- Video
- Audio
- Review
- Inquiry
- WebQuest
- Assessment
- Concepts in Motion
- Multilingual eGlossary

Lesson 1

Reading Guide

Key Concepts
ESSENTIAL QUESTIONS

- How do fossils form?
- How do scientists date fossils?
- How are fossils evidence of biological evolution?

Vocabulary

fossil record p. 189
mold p. 191
cast p. 191
trace fossil p. 191
geologic time scale p. 193
extinction p. 194
biological evolution p. 195

 Multilingual eGlossary

 BrainPOP®

Fossil Evidence of Evolution

Inquiry **What can be learned from fossils?**

When scientists find fossils, they use them as evidence to try to answer questions about past life on Earth. When did this organism live? What did this organism eat? How did it move or grow? How did this organism die? To what other organisms is this one related?

188 • Chapter 6
ENGAGE

Inquiry Launch Lab

20 minutes

How do fossils form?

Evidence from fossils helps scientists understand how organisms have changed over time. Some fossils form when impressions left by organisms in sand or mud are filled in by sediments that harden.

1. Read and complete a lab safety form.
2. Place a **container of moist sand** on top of **newspaper**. Press a **shell** into the moist sand. Carefully remove the shell. Brush any sand on the shell onto the newspaper.
3. Observe the impression, and record your observations in your Science Journal.
4. Pour **plaster of paris** into the impression. Wait for it to harden.
⚠ The mix gets hot as it sets—do not touch it until it has hardened.
5. Remove the shell fossil from the sand, and brush it off.
6. Observe the structure of the fossil, and record your observations.

Think About This

1. What effect did the shell have on the sand?
2. 🔑 **Key Concept** What information do you think someone could learn about the shell and the organism that lived inside it by examining the fossil?

The Fossil Record

On your way to school, you might have seen an oak tree or heard a robin. Although these organisms shed leaves or feathers, their characteristics remain the same from day to day. It might seem as if they have been on Earth forever. However, if you were to travel a few million years back in time, you would not see oak trees or robins. You would see different species of trees and birds. That is because species change over time.

You might already know that fossils are the remains or evidence of once-living organisms. *The* **fossil record** *is made up of all the fossils ever discovered on Earth.* It contains millions of fossils that represent many thousands of species. Most of these species are no longer alive on Earth. The fossil record provides evidence that species have changed over time. Fossils help scientists picture what these species looked like. **Figure 1** shows how scientists think the giant bird *Titanus* might have looked when it was alive. The image is based on fossils that have been discovered and are represented in the photo on the previous page.

The fossil record is enormous, but it is still incomplete. Scientists think it represents only a small fraction of all the organisms that have ever lived on Earth.

Figure 1 Based on fossil evidence, scientists can recreate the physical appearance of species that are no longer alive on Earth.

Fossil Formation

If you have ever seen vultures or other animals eating a dead animal, you know they leave little behind. Any soft tissues animals do not eat, bacteria break down. Only the dead animal's hard parts, such as bones, shells, and teeth, remain. In most instances, these hard parts also break down over time. However, under rare conditions, some become fossils. The soft tissues of animals and plants, such as skin, muscles, or leaves, can also become fossils, but these are even more rare. Some of the ways that fossils can form are shown in **Table 1**.

Reading Check Why is it rare for soft tissue to become a fossil?

Mineralization

After an organism dies, its body could be buried under mud, sand, or other sediments in a stream or river. If minerals in the water replace the organism's original material and harden into rock, a fossil forms. This process is called mineralization. Minerals in water also can filter into the small spaces of a dead organism's tissues and become rock. Most mineralized fossils are of shell or bone, but wood can also become a mineralized fossil, as shown in **Table 1**.

Carbonization

In carbonization, a fossil forms when a dead organism is compressed over time and pressure drives off the organism's liquids and gases. As shown in **Table 1**, only the carbon outline, or film, of the organism remains.

SCIENCE USE V. COMMON USE

tissue
Science Use similar cells that work together and perform a function

Common Use a piece of soft, absorbent paper

Table 1 Fossils form in several ways.

Visual Check What types of organisms or tissues are often preserved as carbon films?

Table 1 How Fossils Form		
	Mineralization	**Carbonization**
Description	Rock-forming minerals, such as calcium carbonate ($CaCO_3$), in water filled in the small spaces in the tissue of these pieces of petrified wood. Water also replaced some of the wood's tissue. Mineralization can preserve the internal structures of an organism.	Fossil films made by carbonization are usually black or dark brown. Fish, insects, and plant leaves, such as this fern frond, are often preserved as carbon films.
Example		

Molds and Casts

Sometimes when an organism dies, its shell or bone might make an impression in mud or sand. When the sediment hardens, so does the impression. *The impression of an organism in a rock is called a* **mold**. Sediments can later fill in the mold and harden to form a cast. *A* **cast** *is a fossil copy of an organism in a rock.* A single organism can form both a mold and a cast, as shown in Table 1. Molds and casts show only external features of organisms.

Trace Fossils

Evidence of an organism's movement or behavior—not just its physical structure—also can be preserved in rock. *A* **trace fossil** *is the preserved evidence of the activity of an organism.* For example, an organism might walk across mud. The tracks, such as the ones shown in Table 1, can fossilize if they are filled with sediment that hardens.

Original Material

In rare cases, the original tissues of an organism can be preserved. Examples of original-material fossils include mammoths frozen in ice and saber-toothed cats preserved in tar pits. Fossilized remains of ancient humans have been found in bogs. Most of these fossils are younger than 10,000 years old. However, the insect encased in amber in Table 1 is millions of years old. Scientists also have found original tissue preserved in the bone of a dinosaur that lived 70 million years ago (mya).

Key Concept Check List the different ways fossils can form.

WORD ORIGIN

fossil
from Latin *fossilis*, means "to obtain by digging"

Molds and Casts	Trace Fossils	Original Material
When sediments hardened around this buried trilobite, a mold formed. Molds are usually of hard parts, such as shells or bone. If a mold is later filled with more sediments that harden, the mold can form a cast.	These footprints were made when a dinosaur walked across mud that later hardened. This trace fossil might provide evidence of the speed and weight of the dinosaur.	If original tissues of organisms are buried in the absence of oxygen for long periods of time, they can fossilize. The insect in this amber became stuck in tree sap that later hardened.

Dating Fossils

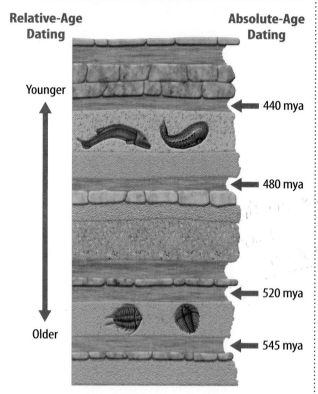

Figure 2 If the age of the igneous layers is known, as shown above, it is possible to estimate the age of the sedimentary layers—and the fossils they contain—between them.

Visual Check What is the estimated age of the trilobite fossils (bottom layer of fossils)?

REVIEW VOCABULARY
isotopes
atoms of the same element that have different numbers of neutrons

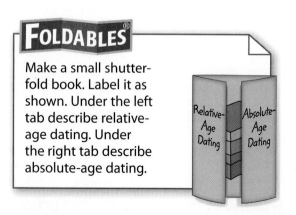

Make a small shutter-fold book. Label it as shown. Under the left tab describe relative-age dating. Under the right tab describe absolute-age dating.

Determining a Fossil's Age

Scientists cannot date most fossils directly. Instead, they date the rocks the fossils are embedded inside. Rocks erode or are recycled over time. However, scientists can determine ages for most of Earth's rocks.

Relative-Age Dating

How does your age compare to the ages of those around you? You might be younger than a brother but older than a sister. This is your relative age. Similarly, a rock is either older or younger than rocks nearby. In relative-age dating, scientists determine the relative order in which rock layers were deposited. In an undisturbed rock formation, they know that the bottom layers are oldest and the top layers are youngest, as shown in **Figure 2**. Relative-age dating helps scientists determine the relative order in which species have appeared on Earth over time.

 Key Concept Check How does relative-age dating help scientists learn about fossils?

Absolute-Age Dating

Absolute-age dating is more precise than relative-age dating. Scientists take advantage of radioactive decay, a natural clocklike process in rocks, to learn a rock's absolute age, or its age in years. In radioactive decay, unstable isotopes in rocks change into stable isotopes over time. Scientists measure the ratio of unstable isotopes to stable isotopes to find the age of a rock. This ratio is best measured in igneous rocks.

Igneous rocks form from volcanic magma. Magma is so hot that it is rare for parts of organisms in it to remain and form fossils. Most fossils form in sediments, which become sedimentary rock. To measure the age of sedimentary rock layers, scientists calculate the ages of igneous layers above and below them. In this way, they can estimate the ages of the fossils embedded within the sedimentary layers, as shown in **Figure 2**.

Fossils over Time

How old do you think Earth's oldest fossils are? You might be surprised to learn that evidence of microscopic, unicellular organisms has been found in rocks 3.4 billion years old. The oldest fossils visible to the unaided eye are about 565 million years old.

The Geologic Time Scale

It is hard to keep track of time that is millions and billions of years long. Scientists organize Earth's history into a time line called the geologic time scale. *The **geologic time scale** is a chart that divides Earth's history into different time units.* The longest time units in the geological time scale are eons. As shown in **Figure 3**, Earth's history is divided into four eons. Earth's most recent eon—the Phanerozoic (fa nuh ruh ZOH ihk) eon—is subdivided into three eras, also shown in **Figure 3**.

 Reading Check What is the geologic time scale?

Dividing Time

You might have noticed in **Figure 3** that neither eons nor eras are equal in length. When scientists began developing the geologic time scale in the 1800s, they did not have absolute-age dating methods. To mark time boundaries, they used fossils. Fossils provided an easy way to mark time. Scientists knew that different rock layers contained different types of fossils. Some of the fossils scientists use to mark the time boundaries are shown in **Figure 3**.

Often, a type of fossil found in one rock layer did not appear in layers above it. Even more surprising, entire collections of fossils in one layer were sometimes absent from layers above them. It seemed as if whole communities of organisms had suddenly disappeared.

 Reading Check What do scientists use to mark boundaries in the geologic time scale?

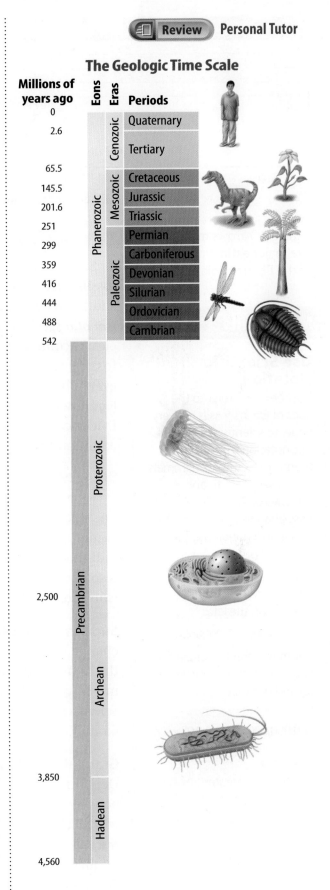

Figure 3 The Phanerozoic eon began about 540 million years ago and continues to the present day. It contains most of Earth's fossil record.

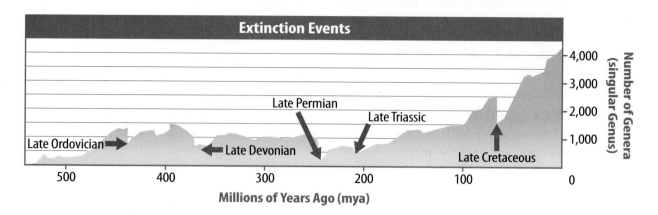

Figure 4 Arrows mark the five major extinction events of the Phanerozoic eon.

Math Skills

Use Scientific Notation

Numbers that refer to the ages of Earth's fossils are very large, so scientists use scientific notation to work with them. For example, mammals appeared on Earth about 200 mya or 200,000,000 years ago. Change this number to scientific notation using the following process.

Move the decimal point until only one nonzero digit remains on the left.

200,000,000 = 2.00000000

Count the number of places you moved the decimal point (8) and use that number as a power of ten.

$200{,}000{,}000 = 2.0 \times 10^8$ years.

Practice

The first vertebrates appeared on Earth about 490,000,000 years ago. Express this time in scientific notation.

- Review
- Math Practice
- Personal Tutor

Extinctions

Scientists now understand that sudden disappearances of fossils in rock layers are evidence of extinction (ihk STINGK shun) events. **Extinction** *occurs when the last individual organism of a species dies.* A mass extinction occurs when many species become extinct within a few million years or less. The fossil record contains evidence that five mass extinction events have occurred during the Phanerozoic eon, as shown in **Figure 4**. Extinctions also occur at other times, on smaller scales. Evidence from the fossil record suggests extinctions have been common throughout Earth's history.

Environmental Change

What causes extinctions? Populations of organisms depend on resources in their environment for food and shelter. Sometimes environments change. After a change happens, individual organisms of a species might not be able to find the resources they need to survive. When this happens, the organisms die, and the species becomes extinct.

Sudden Changes Extinctions can occur when environments change quickly. A volcanic eruption or a meteorite impact can throw ash and dust into the atmosphere, blocking sunlight for many years. This can affect global climate and food webs. Scientists hypothesize that the impact of a huge meteorite 65 million years ago contributed to the extinction of dinosaurs.

Gradual Changes Not all environmental change is sudden. Depending on the location, Earth's tectonic plates move between 1 and 15 cm each year. As plates move and collide with each other over time, mountains form and oceans develop. If a mountain range or an ocean isolates a species, the species might become extinct if it cannot find the resources it needs. Species also might become extinct if sea level changes.

Reading Check What is the relationship between extinction and environmental change?

Extinctions and Evolution

The fossil record contains clear evidence of the extinction of species over time. But it also contains evidence of the appearance of many new species. How do new species arise?

Many early scientists thought that each species appeared on Earth independently of every other species. However, as more fossils were discovered, patterns in the fossil record began to emerge. Many fossil species in nearby rock layers had similar body plans and similar structures. It appeared as if they were related. For example, the series of horse fossils in **Figure 5** suggests that the modern horse is related to other extinct species. These species changed over time in what appeared to be a sequence. Change over time is evolution. **Biological evolution** *is the change over time in populations of related organisms.* Charles Darwin developed a theory about how species evolve from other species. You will read about Darwin's theory in the next lesson.

 Key Concept Check How are fossils evidence of biological evolution?

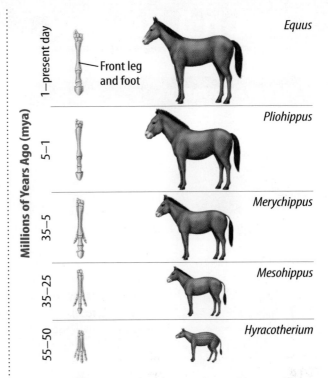

Figure 5 The fossil record is evidence that horses descended from organisms for which only fossils exist today.

Inquiry MiniLab

20 minutes

How do species change over time?

Over long time periods on Earth, certain individuals within populations of organisms were able to survive better than others.

1. Choose a species from the **Species I.D. Cards.**
2. On **chart paper,** draw six squares in a row and number them 1–6, respectively. Use **colored pencils** and **markers** to make a comic strip showing the ancestral and present-day forms of your species in frames 1 and 6.
3. Use information from the I.D. Card to show what you think would be the progression of changes in the species in frames 2–5.
4. In speech bubbles, explain how each change helped the species to survive.

Analyze and Conclude

1. **Infer** why a scientist would identify a fossil from the species in the first frame of your cartoon as the ancestral form of the present-day species.
2. **Key Concept** How would the fossils of the species at each stage provide evidence of biological change over time?

Lesson 1 Review

Visual Summary

Fossils can consist of the hard parts or soft parts of organisms. Fossils can be an impression of an organism or consist of original tissues.

Scientists determine the age of a fossil through relative-age dating or absolute-age dating.

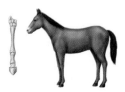

Scientists use fossils as evidence that species have changed over time.

FOLDABLES

Use your lesson Foldable to review the lesson. Save your Foldable for the project at the end of the chapter.

What do you think NOW?

You first read the statements below at the beginning of the chapter.

1. Original tissues can be preserved as fossils.
2. Organisms become extinct only in mass extinction events.

Did you change your mind about whether you agree or disagree with the statements? Rewrite any false statements to make them true.

Use Vocabulary

1. All of the fossils ever found on Earth make up the _____.

2. When the last individual of a species dies, _____ occurs.

3. **Use the term** *biological evolution* in a sentence.

Understand Key Concepts

4. Which is the preserved evidence of the activity of an organism?
   A. cast
   B. mold
   C. fossil film
   D. trace fossil

5. **Explain** why the hard parts of organisms fossilize more often than soft parts.

6. **Draw and label** a diagram that shows how scientists date sedimentary rock layers.

Interpret Graphics

7. **Identify** Copy and fill in the table below to provide examples of changes that might lead to an extinction event.

Sudden changes	
Gradual changes	

Critical Thinking

8. **Infer** If the rock layers shown below have not been disturbed, what type of dating method would help you determine which layer is oldest? Explain.

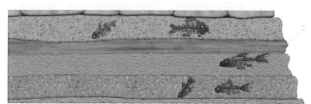

Math Skills

9. Dinosaurs disappeared from Earth about 65,000,000 years ago. Express this number in scientific notation.

Inquiry Skill Practice: Observe

40 minutes

Can you observe changes through time in collections of everyday objects?

Materials

picture sets of items that have changed over time

Everyday objects that are invented, designed, and manufactured by humans often exhibit changes over time in both structure and function. How have these changes affected the efficiency and/or safety of some common items?

Learn It

When scientists **observe** phenomena, they use their senses, such as sight, hearing, touch, and smell. They examine the entire object or situation first, then look carefully for details. After completing their observations, scientists use words or numbers to describe what they saw.

Try It

1. Working with your group members, choose a set of items that you wish to observe, such as telephones, bicycles, or automobiles.

2. Examine the pictures and observe how the item has changed over time.

3. Record your observations in your Science Journal.

4. Observe details of the structure and function of each of the items. Record your observations.

Apply It

5. **Present** your results in the form of an illustrated time line, a consumer magazine article, a role-play of a person-on-the-street interview, a television advertisement, or an idea of your own approved by your teacher.

6. **Key Concept** Identify how your product changed over time and in what ways the changes affected the efficiency and/or safety of the product.

Lesson 1 EXTEND • 197

Lesson 2

Theory of Evolution by Natural Selection

Reading Guide

Key Concepts 🗝
ESSENTIAL QUESTIONS

- Who was Charles Darwin?
- How does Darwin's theory of evolution by natural selection explain how species change over time?
- How are adaptations evidence of natural selection?

Vocabulary

naturalist p. 199
variation p. 201
natural selection p. 202
adaptation p. 203
camouflage p. 204
mimicry p. 204
selective breeding p. 205

 Multilingual eGlossary

 Video

What's Science Got to do With It?

Inquiry Are these exactly the same?

Look closely at these zebras. Are they all exactly the same? How are they different? What accounts for these differences? How do the stripes help these organisms survive in their environment?

198 • Chapter 6
ENGAGE

Launch Lab

20 minutes

Are there variations within your class?

All populations contain variations in some characteristics of their members.

1. Read and complete a lab safety form.
2. Use a **meterstick** to measure the length from your elbow to the tip of your middle finger in centimeters. Record the measurement in your Science Journal.
3. Add your measurement to the class list.
4. Organize all of the measurements from shortest to longest.
5. Break the data into regular increments, such as 31–35 cm, 36–40 cm, and 41–45 cm. Count the number of measurements within each increment.
6. Construct a bar graph using the data. Label each axis and give your graph a title.

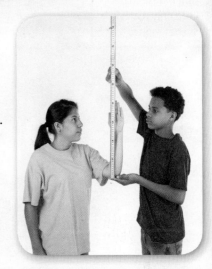

Think About This

1. What are the shortest and longest measurements?
2. How much do the shortest and longest lengths vary from each other?
3. **Key Concept** Describe how your results provide evidence of variations within your classroom population.

Charles Darwin

How many species of birds can you name? You might think of robins, penguins, or even chickens. Scientists estimate that about 10,000 species of birds live on Earth today. Each bird species has similar characteristics. Each has wings, feathers, and a beak. Scientists hypothesize that all birds evolved from an earlier, or ancestral, population of birdlike organisms. As this population evolved into different species, birds became different sizes and colors. They developed different songs and eating habits, but all retained similar bird characteristics.

How do birds and other species evolve? One scientist who worked to answer this question was Charles Darwin. Darwin was an English naturalist who, in the mid-1800s, developed a theory of how evolution works. A **naturalist** is *a person who studies plants and animals by observing them.* Darwin spent many years observing plants and animals in their natural habitats before developing his theory. Recall that a theory is an explanation of the natural world that is well supported by evidence. Darwin was not the first to develop a theory of evolution, but his theory is the one best supported by evidence today.

Key Concept Check Who was Charles Darwin?

FOLDABLES

Make a small four-door shutterfold book. Use it to investigate the who, what, when, and where of Charles Darwin, the Galápagos Islands, and the theory of evolution by natural selection.

Lesson 2
EXPLORE

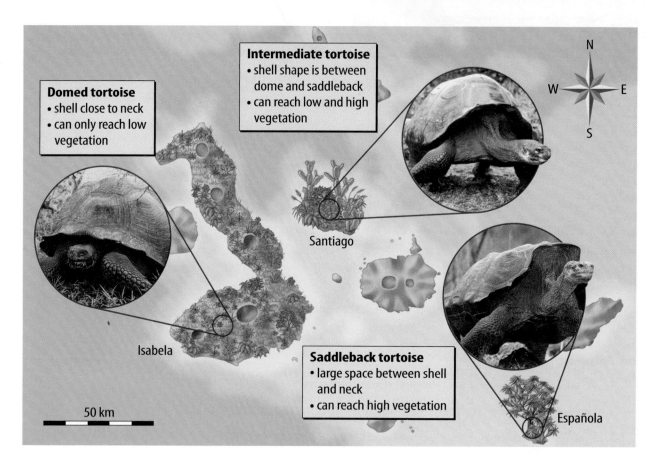

Figure 6 Each island in the Galápagos has a different environment. Tortoises look different depending on which island environment they inhabit.

Visual Check What type of vegetation do domed tortoises eat?

Voyage of the *Beagle*

Darwin served as a naturalist on the HMS *Beagle,* a survey ship of the British navy. During his voyage around the world, Darwin observed and collected many plants and animals.

The Galápagos Islands

Darwin was especially interested in the organisms he saw on the Galápagos (guh LAH puh gus) Islands. The islands, shown in **Figure 6,** are located 1,000 km off the South American coast in the Pacific Ocean. Darwin saw that each island had a slightly different environment. Some were dry. Some were more humid. Others had mixed environments.

Tortoises Giant tortoises lived on many of the islands. When a resident told him that the tortoises on each island looked different, as shown in **Figure 6,** Darwin became curious.

Mockingbirds and Finches Darwin also became curious about the variety of mockingbirds and finches he saw and collected on the islands. Like the tortoises, different types of mockingbirds and finches lived in different island environments. Later, he was surprised to learn that many of these varieties were different enough to be separate species.

Reading Check What made Darwin become curious about the organisms that lived on the Galápagos Islands?

Darwin's Theory

Darwin realized there was a relationship between each species and the food sources of the island it lived on. Look again at **Figure 6.** You can see that tortoises with long necks lived on islands that had tall cacti. Their long necks enabled them to reach high to eat the cacti. The tortoises with short necks lived on islands that had plenty of short grass.

Common Ancestors

Darwin became **convinced** that all the tortoise species were related. He thought they all shared a common ancestor. He suspected that a storm had carried a small ancestral tortoise population to one of the islands from South America millions of years before. Eventually, the tortoises spread to the other islands. Their neck lengths and shell shapes changed to match their islands' food sources. How did this happen?

Variations

Darwin knew that individual members of a species exhibit slight differences, or variations. *A* **variation** *is a slight difference in an inherited trait of individual members of a species.* Even though the snail shells in **Figure 7** are not all exactly the same, they are all from snails of the same species. You can also see variations in the zebras in the photo at the beginning of this lesson. Variations arise naturally in populations. They occur in the offspring as a result of sexual reproduction. You might recall that variations are caused by random mutations, or changes, in genes. Mutations can lead to changes in phenotype. Recall that an organism's phenotype is all of the observable traits and characteristics of the organism. Genetic changes to phenotype can be passed on to future generations.

ACADEMIC VOCABULARY
convince
(verb) to overcome by argument

Figure 7 The variations among the shells of a species of tree snail occur naturally within the population.

Visual Check Describe three variations among these snail shells.

Natural Selection

Darwin did not know about genes. But he realized that variations were the key to the puzzle of how populations of tortoises and other organisms evolved. Darwin understood that food is a limiting resource, which means that the food in each island environment could not support every tortoise that was born. Tortoises had to compete with each other for food. As the tortoises spread to the various islands, some were born with random variations in neck length. If a variation benefited a tortoise, allowing it to compete for food better than other tortoises, the tortoise lived longer. Because it lived longer, it reproduced more. It passed on its variations to its offspring.

This is Darwin's theory of evolution by natural selection. **Natural selection** *is the process by which populations of organisms with variations that help them survive in their environments live longer, compete better, and reproduce more than those that do not have the variations.* Natural selection explains how populations change as their environments change. It explains the process by which Galápagos tortoises became matched to their food sources, as illustrated in **Figure 8**. It also explains the diversity of the Galápagos finches and mockingbirds. Birds with beak variations that help them compete for food live longer and reproduce more.

Key Concept Check What role do variations have in the theory of evolution by natural selection?

Natural Selection

Review Personal Tutor

❶ Reproduction
A population of tortoises produces many offspring that inherit its characteristics.

❷ Variation
A tortoise is born with a variation that makes its neck slightly longer.

❸ Competition
Due to limited resources, not all offspring will survive. An offspring with a longer neck can eat more cacti than other tortoises. It lives longer and produces more offspring.

❹ Selection
Over time, the variation is inherited by more and more offspring. Eventually, all tortoises have longer necks.

Figure 8 A beneficial variation in neck length spreads through a tortoise population by natural selection.

Adaptations

Natural selection explains how all species change over time as their environments change. Through natural selection, a helpful variation in one individual can be passed on to future members of a population. As time passes, more variations arise. The accumulation of many similar variations can lead to an adaptation (a dap TAY shun). *An* **adaptation** *is an inherited trait that increases an organism's chance of surviving and reproducing in its environment.* The long neck of certain species of tortoises is an adaptation to an environment with tall cacti.

 Key Concept Check How do variations lead to adaptations?

Types of Adaptations

Every species has many adaptations. Scientists classify adaptations into three categories: structural, behavioral, and functional. Structural adaptations involve color, shape, and other physical characteristics. The shape of a tortoise's neck is a structural adaptation. Behavioral adaptations involve the way an organism behaves or acts. Hunting at night and moving in herds are examples of behavioral adaptations. Functional adaptations involve internal body systems that affect biochemistry. A drop in body temperature during hibernation is an example of a functional adaptation. **Figure 9** illustrates examples of all three types of adaptations in the desert jackrabbit.

WORD ORIGIN
adaptation
from Latin *adaptare*, means "to fit"

REVIEW VOCABULARY
biochemistry
the study of chemical processes in living organisms

Figure 9 The desert jackrabbit has structural, behavioral, and functional adaptations. These adaptations enable it to survive in its desert environment.

Structural adaptation The jackrabbit's powerful legs help it run fast to escape from predators.

Behavioral adaptation The jackrabbit stays still during the hottest part of the day, helping it conserve energy.

Functional adaptation The blood vessels in the jackrabbit's ears expand to enable the blood to cool before re-entering the body.

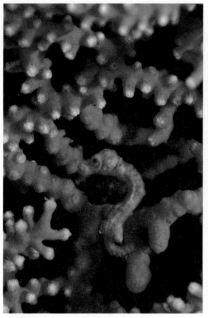
Seahorse

Caterpillar

Pelican

▲ **Figure 10** 🗝
Species evolve adaptations as they interact with their environments, which include other species.

Figure 11 This orchid and its moth pollinator have evolved so closely together that one cannot exist without the other. ▼

Environmental Interactions

Have you ever wanted to be invisible? Many species have evolved adaptations that make them nearly invisible. The seahorse in **Figure 10** is the same color and has a texture similar to the coral it is resting on. This is a structural adaptation called camouflage (KAM uh flahj). **Camouflage** *is an adaptation that enables a species to blend in with its environment.*

Some species have adaptations that draw attention to them. The caterpillar in **Figure 10** resembles a snake. Predators see it and are scared away. *The resemblance of one species to another species is* **mimicry** (MIH mih kree). Camouflage and mimicry are adaptations that help species avoid being eaten. Many other adaptations help species eat. The pelican in **Figure 10** has a beak and mouth uniquely adapted to its food source—fish.

 Reading Check How do camouflage and mimicry differ?

Environments are complex. Species must adapt to an environment's living parts as well as to an environment's nonliving parts. Nonliving things include temperature, water, nutrients in soil, and climate. Deciduous trees shed their leaves due to changes in climate. Camouflage, mimicry, and mouth shape are adaptations mostly to an environment's living parts. An extreme example of two species adapting to each other is shown in **Figure 11**.

Living and nonliving factors are always changing. Even slight environmental changes affect how species adapt. If a species is unable to adapt, it becomes extinct. The fossil record contains many fossils of species unable to adapt to change.

Artificial Selection

Adaptations provide evidence of how closely Earth's species match their environments. This is exactly what Darwin's theory of evolution by natural selection predicted. Darwin provided many examples of adaptation in *On the Origin of Species*, the book he wrote to explain his theory. Darwin did not write this book until 20 years after he developed his theory. He spent those years collecting more evidence for his theory by studying barnacles, orchids, corals, and earthworms.

Darwin also had a hobby of breeding domestic pigeons. He selectively bred pigeons of different colors and shapes to produce new, fancy varieties. *The breeding of organisms for desired characteristics is called* **selective breeding.** Like many domestic plants and animals produced from selective breeding, pigeons look different from their ancestors, as shown in **Figure 12**. Darwin realized that changes caused by selective breeding were much like changes caused by natural selection. Instead of nature selecting variations, humans selected them. Darwin called this process artificial selection.

Artificial selection explains and supports Darwin's theory. As you will read in Lesson 3, other evidence also supports the idea that species evolve from other species.

Figure 12 The pouter pigeon (bottom left) and the fantail pigeon (bottom right) were derived from the wild rock pigeon (top).

Inquiry MiniLab

20 minutes

Who survives?

Camouflage helps organisms blend in. This can help them avoid predators or sneak up on prey. Camouflage helps organisms survive in their environments.

1. Read and complete a lab safety form.
2. Choose an area of your classroom where your moth will rest with open wings during the day.
3. Use **scissors, paper, markers,** and a **ruler** to design a moth that measures 2–5 cm in width with open wings and will be camouflaged where it is placed. Write the location where the moth is to be placed. Give the location and your completed moth to your teacher.
4. On the following day, you will have 1 minute to spot as many moths in the room as you can.
5. In your Science Journal, record the location of moths spotted by your team.
6. Find the remaining moths that were not spotted. Observe their appearance.

Analyze and Conclude

1. **Compare** the appearances and resting places of the moths that were spotted with those that were not spotted.
2. **Key Concept** Explain how camouflage enables an organism to survive in its environment.

Lesson 2 Review

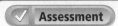 Assessment   Online Quiz

Visual Summary

Charles Darwin developed his theory of evolution partly by observing organisms in their natural environments.

Natural selection occurs when organisms with certain variations live longer, compete better, and reproduce more often than organisms that do not have the variations.

Adaptations occur when a beneficial variation is eventually inherited by all members of a population.

FOLDABLES

Use your lesson Foldable to review the lesson. Save your Foldable for the project at the end of the chapter.

What do you think NOW?

You first read the statements below at the beginning of the chapter.

3. Environmental change causes variations in populations.

4. Variations can lead to adaptations.

Did you change your mind about whether you agree or disagree with the statements? Rewrite any false statements to make them true.

Use Vocabulary

1. A person who studies plants and animals by observing them is a(n) _____.

2. Through _____, populations of organisms adapt to their environments.

3. Some species blend in to their environments through _____.

Understand Key Concepts

4. The observation that the Galápagos tortoises did not all live in the same environment helped Darwin
   A. develop his theory of adaptation.
   B. develop his theory of evolution.
   C. observe mimicry in nature.
   D. practice artificial selection.

5. **Assess** the importance of variations to natural selection.

6. **Compare and contrast** natural selection and artificial selection.

Interpret Graphics

7. **Explain** how the shape of the walking stick at right helps the insect survive in its environment.

8. **Sequence** Copy the graphic organizer below and sequence the steps by which a population of organisms changes by natural selection.

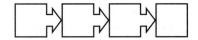

Critical Thinking

9. **Conclude** how Earth's birds developed their diversity through natural selection.

CAREERS in SCIENCE

Peter and Rosemary Grant

Observing Natural Selection

Charles Darwin was a naturalist during the mid-1800s. Based on his observations of nature, he developed the theory of evolution by natural selection. Do scientists still work this way—drawing conclusions from observations? Is there information still to be learned about natural selection? The answer to both questions is yes.

Peter and Rosemary Grant are naturalists who have observed finches in the Galápagos Islands for more than 30 years. They have found that variations in the finches' food supply determine which birds will survive and reproduce. They have observed natural selection in action.

The Grants live on Daphne Major, an island in the Galápagos, for part of each year. They observe and take measurements to compare the size and shape of finches' beaks from year to year. They also examine the kinds of seeds and nuts available for the birds to eat. They use this information to relate changes in the birds' food supply to changes in the finch species' beaks.

The island's ecosystem is fragile, so the Grants take great care not to change the environment of Daphne Major as they observe the finches. They carefully plan their diet to avoid introducing new plant species to the island. They bring all the freshwater they need to drink, and they wash in the ocean. For the Grants, it's just part of the job. As naturalists, they try to observe without interfering with the habitat in which they are living.

▲ Peter and Rosemary Grant make observations and collect data in the field.

▲ This large ground finch is one of the kinds of birds studied by the Grants.

It's Your Turn

RESEARCH AND REPORT Find out more about careers in evolution, ecology, or population biology. What kind of work is done in the laboratory? What kind of work is done in the field? Write a report to explain your findings.

Lesson 2 EXTEND

Lesson 3

Reading Guide

Key Concepts
ESSENTIAL QUESTIONS

- What evidence from living species supports the theory that species descended from other species over time?
- How are Earth's organisms related?

Vocabulary
comparative anatomy p. 210
homologous structure p. 210
analogous structure p. 211
vestigial structure p. 211
embryology p. 212

g Multilingual eGlossary

Biological Evidence of Evolution

Inquiry Does this bird fly?

Some birds, such as the flightless cormorant above, have wings but cannot fly. Their wings are too small to support their bodies in flight. Why do they still have wings? What can scientists learn about the ancestors of present-day birds that have wings but do not fly?

Inquiry Launch Lab

15 minutes

How is the structure of a spoon related to its function?

Would you eat your morning cereal with a spoon that had holes in it? Is using a teaspoon the most efficient way to serve mashed potatoes and gravy to a large group of people? How about using an extra large spoon, or ladle, to eat soup from a small bowl?

1. Read and complete a lab safety form.
2. In a small group, examine your **set of spoons** and discuss your observations.
3. Sketch or describe the structure of each spoon in your Science Journal. Discuss the purpose that each spoon shape might serve.
4. Label the spoons in your Science Journal with their purposes.

Think About This

1. Describe the similarities and differences among the spoons.
2. If spoons were organisms, what do you think the ancestral spoon would look like?
3. **Key Concept** Explain how three of the spoons have different structures and functions, even though they are related by their similarities.

Evidence for Evolution

Recall the sequence of horse fossils from Lesson 1. The sequence might have suggested to you that horses evolved in a straight line—that one species replaced another in a series of orderly steps. Evolution does not occur this way. The diagram in **Figure 13** shows a more realistic version of horse evolution, which looks more like a bush than a straight line. Different horse species were sometimes alive at the same time. They are related to each other because each descended from a common ancestor.

Living species that are closely related share a close common ancestor. The degree to which species are related depends on how closely in time they diverged, or split, from their common ancestor. Although the fossil record is incomplete, it contains many examples of fossil sequences showing close ancestral relationships. Living species show evidence of common ancestry, too.

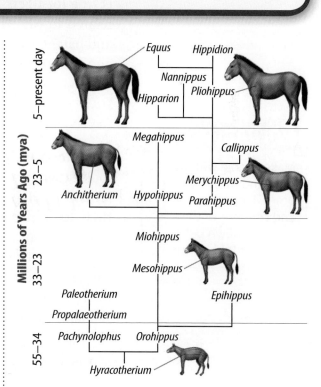

Figure 13 The fossil record indicates that different species of horses often overlapped with each other.

Visual Check Which horse is the common ancestor to all horse species in this graph?

Lesson 3
EXPLORE
209

Homologous Structures

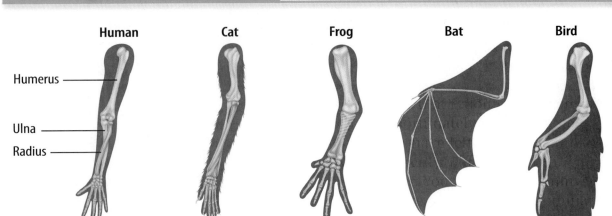

Figure 14 The forelimbs of these species are different sizes, but their placement and structure suggest common ancestry.

Comparative Anatomy

Common ancestry is not difficult to see in many species. For example, it might seem easy to tell that robins, finches, and hawks evolved from a common ancestor. They all have similar features, such as feathers, wings, and beaks. The same is true for tigers, leopards, and house cats. But how are hawks related to cats? How are both hawks and cats related to frogs and bats? Observations of structural and functional similarities and differences in species that do not look alike are possible through comparative anatomy. **Comparative anatomy** *is the study of similarities and differences among structures of living species.*

Homologous Structures Humans, cats, frogs, bats, and birds look different and move in different ways. Humans use their arms for balance and their hands to grasp objects. Cats use their forelimbs to walk, run, and jump. Frogs use their forelimbs to jump. Bats and birds use their forelimbs as wings for flying. However, the forelimb bones of these species exhibit similar patterns, as shown in **Figure 14**. **Homologous** (huh MAH luh gus) **structures** *are body parts of organisms that are similar in structure and position but different in function.*

Homologous structures, such as the forelimbs of humans, cats, frogs, bats, and birds, suggest that these species are related. The more similar two structures are to each other, the more likely it is that the species have evolved from a recent common ancestor.

Key Concept Check How do homologous structures provide evidence for evolution?

FOLDABLES
Make a table with five rows and three columns. Label the rows and columns of the table as shown below. Give your table a title.

	Explanation	Example
Comparative Anatomy		
Vestigial Structures		
Developmental Biology		
Molecular Biology		

Analogous Structures Can you think of a body part in two species that serves the same purpose but differs in structure? How about the wings of birds and flies? Both wings in **Figure 15** are used for flight. But bird wings are covered with feathers. Fly wings are covered with tiny hairs. *Body parts that perform a similar function but differ in structure are* **analogous** (uh NAH luh gus) **structures.** Differences in the structure of bird and fly wings indicate that birds and flies are not closely related.

Vestigial Structures

The bird in the photo at the beginning of this lesson has short, stubby wings. Yet it cannot fly. The bird's wings are an example of vestigial structures. **Vestigial** (veh STIH jee ul) **structures** *are body parts that have lost their original function through evolution.* The best explanation for vestigial structures is that the species with a vestigial structure is related to an ancestral species that used the structure for a specific purpose.

The whale shown in **Figure 16** has tiny pelvic bones inside its body. The presence of pelvic bones in whales suggests that whales descended from ancestors that used legs for walking on land. The fossil evidence supports this conclusion. Many fossils of whale ancestors show a gradual loss of legs over millions of years. They also show, at the same time, that whale ancestors became better adapted to their watery environments.

 Key Concept Check How are vestigial structures evidence of descent from ancestral species?

▲ **Figure 15** Though used for the same function—flight—the wings of birds (top) and insects (bottom) are too different in structure to suggest close common ancestry.

Figure 16 Present-day whales have vestigial structures in the form of small pelvic bones. ▼

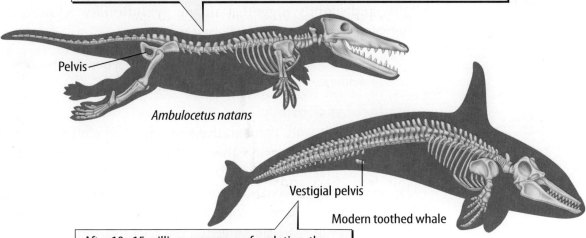

Between 50–40 million years ago, this mammal breathed air and walked clumsily on land. It spent a lot of time in water, but swimming was difficult because of its rear legs. Individuals born with variations that made their rear legs smaller lived longer and reproduced more. This mammal is an ancestor of modern whales.

Pelvis

Ambulocetus natans

Vestigial pelvis

Modern toothed whale

After 10–15 million more years of evolution, the ancestors of modern whales could not walk on land. They were adapted to an aquatic environment. Modern whales have two small vestigial pelvic bones that no longer support legs.

Lesson 3
EXPLAIN

Pharyngeal Pouches

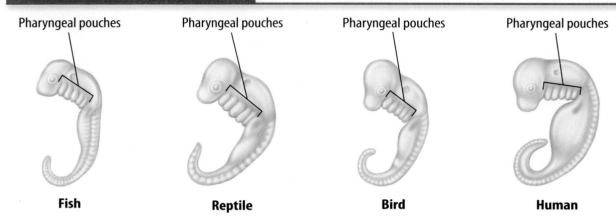

Figure 17 All vertebrate embryos exhibit pharyngeal pouches at a certain stage of their development. These features, which develop into neck and face parts, suggest relatedness.

Developmental Biology

You have just read that studying the internal structures of organisms can help scientists learn more about how organisms are related. Studying the development of embryos can also provide scientists with evidence that certain species are related. *The science of the development of embryos from fertilization to birth is called* **embryology** (em bree AH luh jee).

Pharyngeal Pouches Embryos of different species often resemble each other at different stages of their development. For example, all vertebrate embryos have pharyngeal (fuh rihn JEE ul) pouches at one stage, as shown in **Figure 17.** This feature develops into different body parts in each vertebrate. Yet, in all vertebrates, each part is in the face or neck. For example, in reptiles, birds, and humans, part of the pharyngeal pouch develops into a gland in the neck that regulates calcium. In fish, the same part becomes the gills. One function of gills is to regulate calcium. The similarities in function and location of gills and glands suggest a strong evolutionary relationship between fish and other vertebrates.

 Key Concept Check How do pharyngeal pouches provide evidence of relationships among species?

Molecular Biology

Studies of fossils, comparative anatomy, and embryology provide support for Darwin's theory of evolution by natural selection. Molecular biology is the study of gene structure and function. Discoveries in molecular biology have confirmed and extended much of the data already collected about the theory of evolution. Darwin did not know about genes, but scientists today know that mutations in genes are the source of variations upon which natural selection acts. Genes provide powerful support for evolution.

 Reading Check What is molecular biology?

WORD ORIGIN
embryology
from Greek *embryon,* means "to swell" and from Greek *logia,* means "study of"

Chapter 6
EXPLAIN

Comparing Sequences All organisms on Earth have genes. All genes are made of DNA, and all genes work in similar ways. This supports the idea that all organisms are related. Scientists can study relatedness of organisms by comparing genes and proteins among living species. For example, nearly all organisms contain a gene that codes for cytochrome c, a protein required for cellular respiration. Some species, such as humans and rhesus monkeys, have nearly identical cytochrome c. The more closely related two species are, the more similar their genes and proteins are.

 Key Concept Check How is molecular biology used to determine relationships among species?

Divergence Scientists have found that some stretches of shared DNA mutate at regular, predictable rates. Scientists use this "molecular clock" to estimate at what time in the past living species diverged from common ancestors. For example, as shown in **Figure 18,** molecular data indicate that whales and porpoises are more closely related to hippopotamuses than they are to any other living species.

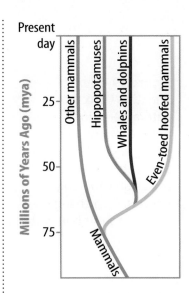

Figure 18 Whales and hippopotamuses share an ancestor that lived 50–60 mya.

Inquiry MiniLab

10 minutes

How related are organisms?

Proteins, such as cytochrome c, are made from combinations of just 20 amino acids. The graph below shows the number of amino acid differences in cytochrome c between humans and other organisms.

1. Use the graph at right to answer the questions below.

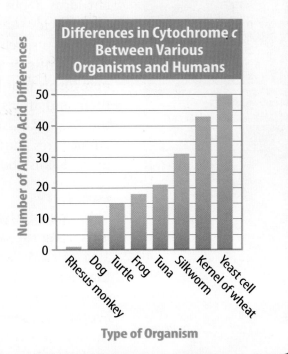

Analyze and Conclude

1. **Identify** Which organism has the least difference in the number of amino acids in cytochrome c compared to humans? Which organism has the most difference?

2. **Infer** Which organisms do you think might be more closely related to each other: a dog and a turtle or a dog and a silkworm? Explain your answer.

3. **Key Concept** Notice the differences in the number of amino acids in cytochrome c between each organism and humans. How might these differences explain the relatedness of each organism to humans?

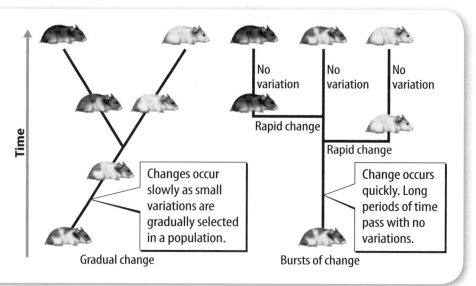

Figure 19 Many scientists think that natural selection produces new species slowly and steadily. Other scientists think species exist stably for long periods, then change occurs in short bursts. ▶

Figure 20 *Tiktaalik* lived 385–359 mya. Like amphibians, it had wrists and lungs. Like fish, it had fins, gills, and scales. Scientists think it is an intermediate species linking fish and amphibians. ▼

The Study of Evolution Today

The theory of evolution by natural selection is the cornerstone of modern biology. Since Darwin published his theory, scientists have confirmed, refined, and extended Darwin's work. They have observed natural selection in hundreds of living species. Their studies of fossils, anatomy, embryology, and molecular biology have all provided evidence of relatedness among living and extinct species.

How New Species Form

New evidence supporting the theory of evolution by natural selection is discovered nearly every day. But scientists debate some of the details. **Figure 19** shows that scientists have different ideas about the rate at which natural selection produces new species—slowly and gradually or quickly, in bursts. The origin of a species is difficult to study on human time scales. It is also difficult to study in the incomplete fossil record. Yet, new fossils that have features of species that lived both before them and after them are discovered all the time. For example, the *Tiktaalik* fossil shown in **Figure 20** has both fish and amphibian features. Further fossil discoveries will help scientists study more details about the origin of new species.

Diversity

How evolution has produced Earth's wide diversity of organisms using the same basic building blocks—genes—is an active area of study in evolutionary biology. Scientists are finding that genes can be reorganized in simple ways and give rise to dramatic changes in organisms. Though scientists now study evolution at the molecular level, the basic principles of Darwin's theory of evolution by natural selection have remained unchanged for over 150 years.

Lesson 3 Review

 Assessment   Online Quiz

Visual Summary

By comparing the anatomy of organisms and looking for homologous or analogous structures, scientists can determine if organisms had a common ancestor.

Some organisms have vestigial structures, suggesting that they descended from a species that used the structure for a purpose.

Pharyngeal pouches

Human

Scientists use evidence from developmental and molecular biology to help determine if organisms are related.

FOLDABLES

Use your lesson Foldable to review the lesson. Save your Foldable for the project at the end of the chapter.

What do you think NOW?

You first read the statements below at the beginning of the chapter.

5. Living species contain no evidence that they are related to each other.

6. Plants and animals share similar genes.

Did you change your mind about whether you agree or disagree with the statements? Rewrite any false statements to make them true.

Use Vocabulary

1. **Define** *embryology* in your own words.

2. **Distinguish** between a homologous structure and an analogous structure.

3. **Use the term** *vestigial structure* in a complete sentence.

Understand Key Concepts

4. Scientists use molecular biology to determine how two species are related by comparing the genes in one species to genes
   A. in extinct species.   C. in related species.
   B. in human species.   D. in related fossils.

5. **Discuss** how pharyngeal pouches provide evidence for biological evolution.

6. **Explain** Some blind cave salamanders have eyes. How might this be evidence that cave salamanders evolved from sighted ancestors?

Interpret Graphics

7. **Interpret** The wings of a flightless cormorant are an example of which type of structure?

8. **Assess** Copy and fill in the graphic organizer below to identify four areas of study that provide evidence for evolution.

Critical Thinking

9. **Predict** what a fossil that illustrates the evolution of a bird from a reptile might look like.

Inquiry Lab

2 class periods

Model Adaptations in an Organism

Materials

clay

colored pencils

colored markers

toothpicks

construction paper

Also needed: creative construction materials, glue, scissors

Safety

Conditions on our planet have changed since Earth formed over 4.5 billion years ago. Changes in the concentrations of gases in the atmosphere, temperature, and the amount of precipitation make Earth different today from when it first formed. Other events, such as volcanic eruptions, meteorite strikes, tsunamis, or wildfires, can drastically and rapidly change the conditions in certain environments. As you have read, Earth's fossil record provides evidence that, over millions of years, many organisms developed adaptations that enabled them to survive as Earth's environmental conditions changed.

Ask a Question

How do adaptations enable an organism to survive changes in the environment?

Make Observations

1. Read and complete a lab safety form.
2. Obtain Version 1.0 of the organism you will model from your teacher.
3. Your teacher will describe Event 1 that has occurred on Earth while your organism is alive. Use markers and a piece of construction paper to design adaptations to your organism that would enable it to survive the changing conditions that result from Event 1. Label the adapted organism *Version 1.1*.
4. For each event that your teacher describes, design and draw the adaptations that would enable your organism to survive the changing conditions. Label each new organism *Version 1.X*, filling in the *X* with the appropriate version number.
5. Use the materials provided to make a model of the final version of your organism, showing all of the adaptations.

Volcanic eruption

Predation

Form a Hypothesis

6. After reviewing and discussing all of the adaptations of your organism, formulate a hypothesis to explain how physical adaptations help an organism survive changes to the environment.

Test Your Hypothesis

7. Research evidence from the fossil record that shows one adaptation that developed and enabled an organism to survive over time under the conditions of one of the environmental events experienced by your model organism.
8. Record the information in your Science Journal.

Analyze and Conclude

9. **Compare** the adaptations that the different groups gave their organisms to survive each event described by your teacher. What kinds of different structures were created to help each organism survive?
10. **The Big Idea** Describe three variations in human populations that would enable some individuals to survive severe environmental changes.

Communicate Your Results

Present your completed organisms to the class and/or judges of "Ultimate Survivor." Explain the adaptations and the reasoning behind them in either an oral presentation or a demonstration, during which classmates and/or judges will review the models.

Inquiry Extension

Compare the organisms made by groups in your class to the organisms created by groups in other sections. Observe the differences in the adaptations of the organisms. In each section, the events were presented in a different order. How might this have affected the final appearance and characteristics of the different organisms?

Meteorite impact

Lab Tips

☑ Make sure you think of all of the implications of an environmental change event before you decide upon an adaptation.

☑ Decide upon your reasoning for the adaptation before putting the adaptation on your model.

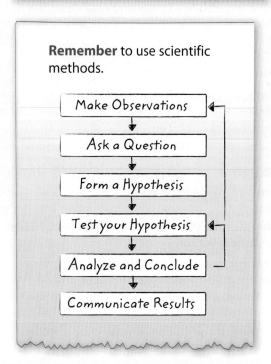

Remember to use scientific methods.

- Make Observations
- Ask a Question
- Form a Hypothesis
- Test your Hypothesis
- Analyze and Conclude
- Communicate Results

Chapter 6 Study Guide

Through natural selection, species evolve as they adapt to Earth's changing environments.

Key Concepts Summary

Lesson 1: Fossil Evidence of Evolution

- Fossils form in many ways, including mineral replacement, carbonization, and impressions in sediment.
- Scientists can learn the ages of fossils by techniques of relative-age dating and absolute-age dating.
- Though incomplete, the **fossil record** contains patterns suggesting the **biological evolution** of related species.

Lesson 2: Theory of Evolution by Natural Selection

- The 19th century **naturalist** Charles Darwin developed a theory of evolution that is still studied today.
- Darwin's theory of evolution by **natural selection** is the process by which populations with **variations** that help them survive in their environments live longer and reproduce more than those without beneficial variations. Over time, beneficial variations spread through populations, and new species that are adapted to their environments evolve.
- **Camouflage, mimicry,** and other **adaptations** are evidence of the close relationships between species and their changing environments.

Lesson 3: Biological Evidence of Evolution

- Fossils provide only one source of evidence of evolution. Additional evidence comes from living species, including studies in **comparative anatomy, embryology,** and molecular biology.
- Through evolution by natural selection, all of Earth's organisms are related. The more recently they share a common ancestor, the more closely they are related.

Vocabulary

fossil record p. 189
mold p. 191
cast p. 191
trace fossil p. 191
geologic time scale p. 193
extinction p. 194
biological evolution p. 195

naturalist p. 199
variation p. 201
natural selection p. 202
adaptation p. 203
camouflage p. 204
mimicry p. 204
selective breeding p. 205

comparative anatomy p. 210
homologous structure p. 210
analogous structure p. 211
vestigial structure p. 211
embryology p. 212

Study Guide

Review
- Personal Tutor
- Vocabulary eGames
- Vocabulary eFlashcards

FOLDABLES Chapter Project

Assemble your lesson Foldables as shown to make a Chapter Project. Use the project to review what you have learned in this chapter.

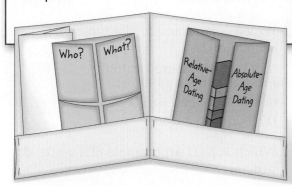

Use Vocabulary

Distinguish between the following terms.

1. *mold* and *cast*
2. *absolute-age dating* and *relative-age dating*
3. *extinction* and *biological evolution*
4. *variations* and *adaptations*
5. *camouflage* and *mimicry*
6. *natural selection* and *selective breeding*
7. *homologous structure* and *analogous structure*
8. *embryology* and *comparative anatomy*
9. *vestigial structure* and *homologous structure*

Link Vocabulary and Key Concepts

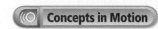 Interactive Concept Map

Copy this concept map, and then use vocabulary terms from the previous page to complete the concept map.

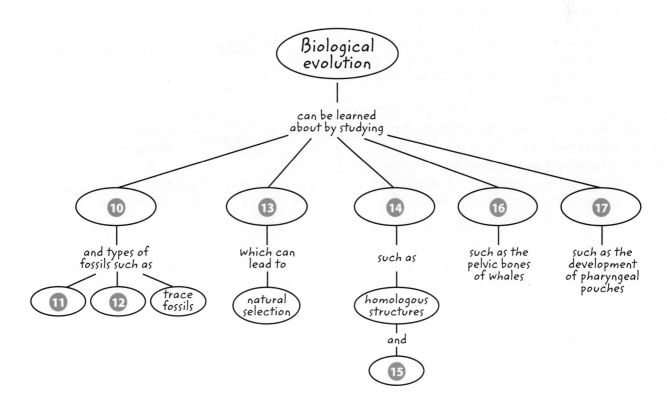

Chapter 6 Review

Understand Key Concepts

1. Why do scientists think the fossil record is incomplete?
   A. Fossils decompose over time.
   B. The formation of fossils is rare.
   C. Only organisms with hard parts become fossils.
   D. There are no fossils before the Phanerozoic eon.

2. What do the arrows on the graph below represent?

   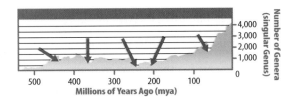

   A. extinction events
   B. meteorite impacts
   C. changes in Earth's temperature
   D. the evolution of a new species

3. What can scientists learn about fossils using techniques of absolute-age dating?
   A. estimated ages of fossils in rock layers
   B. precise ages of fossils in rock layers
   C. causes of fossil disappearances in rock layers
   D. structural similarities to other fossils in rock layers

4. Which is the sequence by which natural selection works?
   A. selection → adaptation → variation
   B. selection → variation → adaptation
   C. variation → adaptation → selection
   D. variation → selection → adaptation

5. Which type of fossil forms through carbonization?
   A. cast
   B. mold
   C. fossil film
   D. trace fossil

6. Which is the source of variations in a population of organisms?
   A. changes in environment
   B. changes in genes
   C. the interaction of genes with an environment
   D. the interaction of individuals with an environment

7. Which is an example of a functional adaptation?
   A. a brightly colored butterfly
   B. birds flying south in the fall
   C. the spray of a skunk
   D. thorns on a rose

8. Which is NOT an example of a vestigial structure?
   A. eyes of a blind salamander
   B. pelvic bones in a whale
   C. thorns on a rose bush
   D. wings on a flightless bird

9. Which do the images below represent?

   Human

   Cat

   A. analogous structures
   B. embryological structures
   C. homologous structures
   D. vestigial structures

10. Which is an example of a sudden change that could lead to the extinction of species?
    A. a mountain range isolates a species
    B. Earth's tectonic plates move
    C. a volcano erupts
    D. sea level changes

Chapter Review

Critical Thinking

11. **Explain** the relationship between fossils and extinction events.

12. **Infer** In 2004, a fossil of an organism that had fins and gills, but also lungs and wrists, was discovered. What might this fossil suggest about evolution?

13. **Summarize** Darwin's theory of natural selection using the Galápagos tortoises or finches as an example.

14. **Assess** how the determination that Earth is 4.6 billion years provided support for the idea that all species evolved from a common ancestor.

15. **Describe** how cytochrome *c* provides evidence of evolution.

16. **Explain** why the discovery of genes was powerful support for Darwin's theory of natural selection.

17. **Interpret Graphics** The diagram below shows two different methods by which evolution by natural selection might proceed. Discuss how these two methods differ.

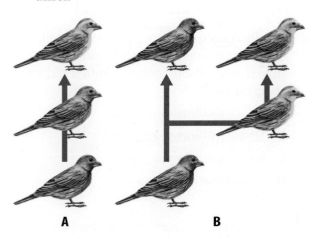

Writing in Science

18. **Write** a paragraph explaining how natural selection and selective breeding are related. Include a main idea, supporting details, and a concluding sentence.

REVIEW THE BIG IDEA

19. How do species adapt to changing environments over time? Explain how evidence from the fossil record and from living species suggests that Earth's species are related. List each type of evidence and provide an example of each.

20. The photo below shows an orchid that looks like a bee. How might this adaptation be evidence of evolution by natural selection?

Math Skills

Use Scientific Notation

21. The earliest fossils appeared about 3,500,000,000 years ago. Express this number in scientific notation.

22. The oldest fossils visible to the unaided eye are about 565,000,000 years old. What is this time in scientific notation?

23. The oldest human fossils are about 1×10^4 years old. Express this as a whole number.

Standardized Test Practice

Record your answers on the answer sheet provided by your teacher or on a sheet of paper.

Multiple Choice

1. Which may form over time from the impression a bird feather makes in mud?
   - A cast
   - B mold
   - C fossil film
   - D trace fossil

2. Which is NOT one of the three main categories of adaptations?
   - A behavioral
   - B functional
   - C pharyngeal
   - D structural

Use the figure below to answer question 3.

Bat wing   Insect wing

3. The figure shows the wings of a bat and an insect. Which term describes these structures?
   - A analogous
   - B developmental
   - C homologous
   - D vestigial

4. What is an adaptation?
   - A a body part that has lost its original function through evolution
   - B a characteristic that better equips an organism to survive in its environment
   - C a feature that appears briefly during early development
   - D a slight difference among the individuals in a species

5. What causes variations to arise in a population?
   - A changes in the environment
   - B competition for limited resources
   - C random mutations in genes
   - D rapid population increases

Use the image below to answer question 6.

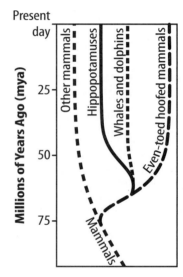

6. The image above shows that even-toed hoofed mammals and other mammals shared a common ancestor. When did this ancestor live?
   - A 25–35 million years ago
   - B 50–60 million years ago
   - C 60–75 million years ago
   - D 75 million years ago

7. Which term describes the method Darwin used that resulted in pigeons with desired traits?
   - A evolution
   - B mimicry
   - C natural selection
   - D selective breeding

Standardized Test Practice

Use the figure below to answer question 8.

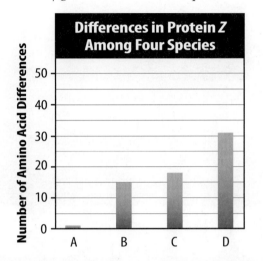

8 The chart shows that species B and C have the fewest amino acid differences for a protein among four species. What does this suggest about their evolutionary relationship?

  A They are more closely related to each other than to the other species.

  B They evolved at a faster rate when compared to the other species.

  C They share a developmental similarity not observed in the other species.

  D They do not share a common ancestor with the other species.

9 Which developmental similarity among all vertebrates is evidence that they share a common ancestor?

  A analogous structures

  B pharyngeal pouches

  C variation rates

  D vestigial structures

Constructed Response

Use the figure below to answer questions 10 and 11.

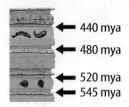

10 What is the approximate age of the fish fossils (top layer of fossils)? Express your answer as a range, and explain how you derived the answer.

11 What type of material or rock most likely forms the layer that contains the fossils? In your response, explain how these fossils formed.

12 Explain how a sudden and drastic environmental change might lead to the extinction of a species.

13 Darwin formulated his theory of evolution by natural selection based on the observation that food is a limiting resource. What did he mean by that? Use the Galápagos tortoises to explain your answer.

14 Explain how the fossil record provides evidence of biological evolution.

NEED EXTRA HELP?														
If You Missed Question...	1	2	3	4	5	6	7	8	9	10	11	12	13	14
Go to Lesson...	1	2	3	2	2	3	2	3	3	1	1	1	2	1

Student Resources

For Students and Parents/Guardians

These resources are designed to help you achieve success in science. You will find useful information on laboratory safety, math skills, and science skills. In addition, science reference materials are found in the Reference Handbook. You'll find the information you need to learn and sharpen your skills in these resources.

Table of Contents

Science Skill Handbook SR-2
Scientific Methods SR-2
- Identify a Question SR-2
- Gather and Organize Information SR-2
- Form a Hypothesis SR-5
- Test the Hypothesis SR-6
- Collect Data SR-6
- Analyze the Data SR-9
- Draw Conclustions SR-10
- Communicate SR-10

Safety Symbols SR-11
Safety in the Science Laboratory SR-12
- General Safety Rules SR-12
- Prevent Accidents SR-12
- Laboratory Work SR-12
- Emergencies SR-13

Math Skill Handbook SR-14
Math Review SR-14
- Use Fractions SR-14
- Use Ratios SR-17
- Use Decimals SR-17
- Use Proportions SR-18
- Use Percentages SR-19
- Solve One-Step Equations SR-19
- Use Statistics SR-20
- Use Geometry SR-21

Science Applications SR-24
- Measure in SI SR-24
- Dimensional Analysis SR-24
- Precision and Significant Digits SR-26
- Scientific Notation SR-26
- Make and Use Graphs SR-27

Reference Handbook SR-29
- Use and Care of a Microscope SR-29
- Diversity of Life: Classification of Living Organisms SR-30
- Periodic Table of the Elements SR-34

Glossary G-2

Index I-2

Credits C-2

Scientific Methods

Scientists use an orderly approach called the scientific method to solve problems. This includes organizing and recording data so others can understand them. Scientists use many variations in this method when they solve problems.

Identify a Question

The first step in a scientific investigation or experiment is to identify a question to be answered or a problem to be solved. For example, you might ask which gasoline is the most efficient.

Gather and Organize Information

After you have identified your question, begin gathering and organizing information. There are many ways to gather information, such as researching in a library, interviewing those knowledgeable about the subject, and testing and working in the laboratory and field. Fieldwork is investigations and observations done outside of a laboratory.

Researching Information Before moving in a new direction, it is important to gather the information that already is known about the subject. Start by asking yourself questions to determine exactly what you need to know. Then you will look for the information in various reference sources, like the student is doing in **Figure 1**. Some sources may include textbooks, encyclopedias, government documents, professional journals, science magazines, and the Internet. Always list the sources of your information.

Figure 1 The Internet can be a valuable research tool.

Evaluate Sources of Information Not all sources of information are reliable. You should evaluate all of your sources of information, and use only those you know to be dependable. For example, if you are researching ways to make homes more energy efficient, a site written by the U.S. Department of Energy would be more reliable than a site written by a company that is trying to sell a new type of weatherproofing material. Also, remember that research always is changing. Consult the most current resources available to you. For example, a 1985 resource about saving energy would not reflect the most recent findings.

Sometimes scientists use data that they did not collect themselves, or conclusions drawn by other researchers. This data must be evaluated carefully. Ask questions about how the data were obtained, if the investigation was carried out properly, and if it has been duplicated exactly with the same results. Would you reach the same conclusion from the data? Only when you have confidence in the data can you believe it is true and feel comfortable using it.

Interpret Scientific Illustrations As you research a topic in science, you will see drawings, diagrams, and photographs to help you understand what you read. Some illustrations are included to help you understand an idea that you can't see easily by yourself, like the tiny particles in an atom in **Figure 2**. A drawing helps many people to remember details more easily and provides examples that clarify difficult concepts or give additional information about the topic you are studying. Most illustrations have labels or a caption to identify or to provide more information.

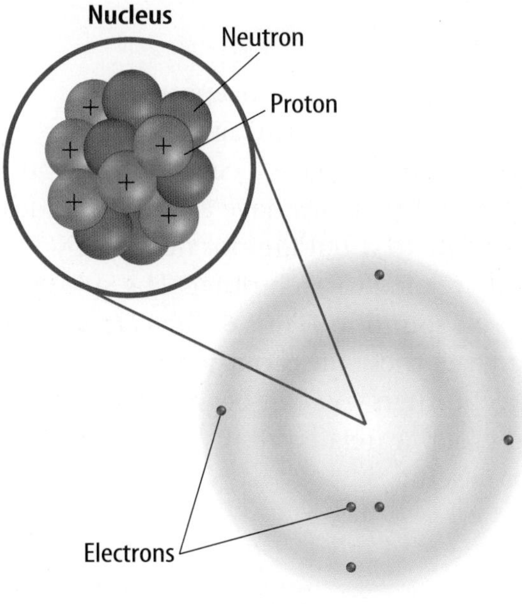

Figure 2 This drawing shows an atom of carbon with its six protons, six neutrons, and six electrons.

Concept Maps One way to organize data is to draw a diagram that shows relationships among ideas (or concepts). A concept map can help make the meanings of ideas and terms more clear, and help you understand and remember what you are studying. Concept maps are useful for breaking large concepts down into smaller parts, making learning easier.

Network Tree A type of concept map that not only shows a relationship, but how the concepts are related is a network tree, shown in **Figure 3**. In a network tree, the words are written in the ovals, while the description of the type of relationship is written across the connecting lines.

When constructing a network tree, write down the topic and all major topics on separate pieces of paper or notecards. Then arrange them in order from general to specific. Branch the related concepts from the major concept and describe the relationship on the connecting line. Continue to more specific concepts until finished.

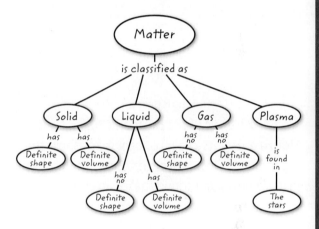

Figure 3 A network tree shows how concepts or objects are related.

Events Chain Another type of concept map is an events chain. Sometimes called a flow chart, it models the order or sequence of items. An events chain can be used to describe a sequence of events, the steps in a procedure, or the stages of a process.

When making an events chain, first find the one event that starts the chain. This event is called the initiating event. Then, find the next event and continue until the outcome is reached, as shown in **Figure 4** one the next page.

Science Skill Handbook • **SR-3**

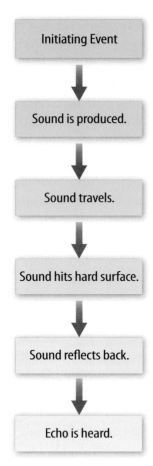

Figure 4 Events-chain concept maps show the order of steps in a process or event. This concept map shows how a sound makes an echo.

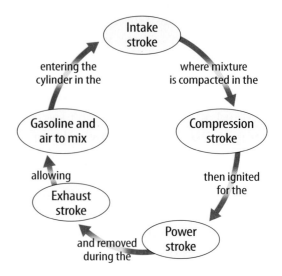

Figure 5 A cycle map shows events that occur in a cycle.

Cycle Map A specific type of events chain is a cycle map. It is used when the series of events do not produce a final outcome, but instead relate back to the beginning event, such as in **Figure 5**. Therefore, the cycle repeats itself.

To make a cycle map, first decide what event is the beginning event. This is also called the initiating event. Then list the next events in the order that they occur, with the last event relating back to the initiating event. Words can be written between the events that describe what happens from one event to the next. The number of events in a cycle map can vary, but usually contain three or more events.

Spider Map A type of concept map that you can use for brainstorming is the spider map. When you have a central idea, you might find that you have a jumble of ideas that relate to it but are not necessarily clearly related to each other. The spider map on sound in **Figure 6** shows that if you write these ideas outside the main concept, then you can begin to separate and group unrelated terms so they become more useful.

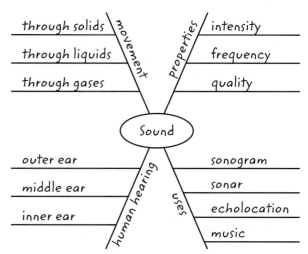

Figure 6 A spider map allows you to list ideas that relate to a central topic but not necessarily to one another.

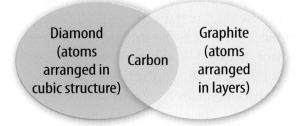

Figure 7 This Venn diagram compares and contrasts two substances made from carbon.

Venn Diagram To illustrate how two subjects compare and contrast you can use a Venn diagram. You can see the characteristics that the subjects have in common and those that they do not, shown in **Figure 7.**

To create a Venn diagram, draw two overlapping ovals that are big enough to write in. List the characteristics unique to one subject in one oval, and the characteristics of the other subject in the other oval. The characteristics in common are listed in the overlapping section.

Make and Use Tables One way to organize information so it is easier to understand is to use a table. Tables can contain numbers, words, or both.

To make a table, list the items to be compared in the first column and the characteristics to be compared in the first row. The title should clearly indicate the content of the table, and the column or row heads should be clear. Notice that in **Table 1** the units are included.

Table 1 Recyclables Collected During Week			
Day of Week	Paper (kg)	Aluminum (kg)	Glass (kg)
Monday	5.0	4.0	12.0
Wednesday	4.0	1.0	10.0
Friday	2.5	2.0	10.0

Make a Model One way to help you better understand the parts of a structure, the way a process works, or to show things too large or small for viewing is to make a model. For example, an atomic model made of a plastic-ball nucleus and chenille-stem electron shells can help you visualize how the parts of an atom relate to each other. Other types of models can be devised on a computer or represented by equations.

Form a Hypothesis

A possible explanation based on previous knowledge and observations is called a hypothesis. After researching gasoline types and recalling previous experiences in your family's car you form a hypothesis—our car runs more efficiently because we use premium gasoline. To be valid, a hypothesis has to be something you can test by using an investigation.

Predict When you apply a hypothesis to a specific situation, you predict something about that situation. A prediction makes a statement in advance, based on prior observation, experience, or scientific reasoning. People use predictions to make everyday decisions. Scientists test predictions by performing investigations. Based on previous observations and experiences, you might form a prediction that cars are more efficient with premium gasoline. The prediction can be tested in an investigation.

Design an Experiment A scientist needs to make many decisions before beginning an investigation. Some of these include: how to carry out the investigation, what steps to follow, how to record the data, and how the investigation will answer the question. It also is important to address any safety concerns.

Science Skill Handbook • **SR-5**

Test the Hypothesis

Now that you have formed your hypothesis, you need to test it. Using an investigation, you will make observations and collect data, or information. This data might either support or not support your hypothesis. Scientists collect and organize data as numbers and descriptions.

Follow a Procedure In order to know what materials to use, as well as how and in what order to use them, you must follow a procedure. **Figure 8** shows a procedure you might follow to test your hypothesis.

Procedure

- **Step 1** Use regular gasoline for two weeks.
- **Step 2** Record the number of kilometers between fill-ups and the amount of gasoline used.
- **Step 3** Switch to premium gasoline for two weeks.
- **Step 4** Record the number of kilometers between fill-ups and the amount of gasoline used.

Figure 8 A procedure tells you what to do step-by-step.

Identify and Manipulate Variables and Controls In any experiment, it is important to keep everything the same except for the item you are testing. The one factor you change is called the independent variable. The change that results is the dependent variable. Make sure you have only one independent variable, to assure yourself of the cause of the changes you observe in the dependent variable. For example, in your gasoline experiment the type of fuel is the independent variable. The dependent variable is the efficiency.

Many experiments also have a control—an individual instance or experimental subject for which the independent variable is not changed. You can then compare the test results to the control results. To design a control you can have two cars of the same type. The control car uses regular gasoline for four weeks. After you are done with the test, you can compare the experimental results to the control results.

Collect Data

Whether you are carrying out an investigation or a short observational experiment, you will collect data, as shown in **Figure 9.** Scientists collect data as numbers and descriptions and organize them in specific ways.

Observe Scientists observe items and events, then record what they see. When they use only words to describe an observation, it is called qualitative data. Scientists' observations also can describe how much there is of something. These observations use numbers, as well as words, in the description and are called quantitative data. For example, if a sample of the element gold is described as being "shiny and very dense" the data are qualitative. Quantitative data on this sample of gold might include "a mass of 30 g and a density of 19.3 g/cm^3."

Figure 9 Collecting data is one way to gather information directly.

Figure 10 Record data neatly and clearly so it is easy to understand.

When you make observations you should examine the entire object or situation first, and then look carefully for details. It is important to record observations accurately and completely. Always record your notes immediately as you make them, so you do not miss details or make a mistake when recording results from memory. Never put unidentified observations on scraps of paper. Instead they should be recorded in a notebook, like the one in **Figure 10.** Write your data neatly so you can easily read it later. At each point in the experiment, record your observations and label them. That way, you will not have to determine what the figures mean when you look at your notes later. Set up any tables that you will need to use ahead of time, so you can record any observations right away. Remember to avoid bias when collecting data by not including personal thoughts when you record observations. Record only what you observe.

Estimate Scientific work also involves estimating. To estimate is to make a judgment about the size or the number of something without measuring or counting. This is important when the number or size of an object or population is too large or too difficult to accurately count or measure.

Sample Scientists may use a sample or a portion of the total number as a type of estimation. To sample is to take a small, representative portion of the objects or organisms of a population for research. By making careful observations or manipulating variables within that portion of the group, information is discovered and conclusions are drawn that might apply to the whole population. A poorly chosen sample can be unrepresentative of the whole. If you were trying to determine the rainfall in an area, it would not be best to take a rainfall sample from under a tree.

Measure You use measurements every day. Scientists also take measurements when collecting data. When taking measurements, it is important to know how to use measuring tools properly. Accuracy also is important.

Length To measure length, the distance between two points, scientists use meters. Smaller measurements might be measured in centimeters or millimeters.

Length is measured using a metric ruler or meterstick. When using a metric ruler, line up the 0-cm mark with the end of the object being measured and read the number of the unit where the object ends. Look at the metric ruler shown in **Figure 11.** The centimeter lines are the long, numbered lines, and the shorter lines are millimeter lines. In this instance, the length would be 4.50 cm.

Figure 11 This metric ruler has centimeter and millimeter divisions.

Science Skill Handbook • **SR-7**

Mass The SI unit for mass is the kilogram (kg). Scientists can measure mass using units formed by adding metric prefixes to the unit gram (g), such as milligram (mg). To measure mass, you might use a triple-beam balance similar to the one shown in **Figure 12**. The balance has a pan on one side and a set of beams on the other side. Each beam has a rider that slides on the beam.

When using a triple-beam balance, place an object on the pan. Slide the largest rider along its beam until the pointer drops below zero. Then move it back one notch. Repeat the process for each rider proceeding from the larger to smaller until the pointer swings an equal distance above and below the zero point. Sum the masses on each beam to find the mass of the object. Move all riders back to zero when finished.

Instead of putting materials directly on the balance, scientists often take a tare of a container. A tare is the mass of a container into which objects or substances are placed for measuring their masses. To find the mass of objects or substances, find the mass of a clean container. Remove the container from the pan, and place the object or substances in the container. Find the mass of the container with the materials in it. Subtract the mass of the empty container from the mass of the filled container to find the mass of the materials you are using.

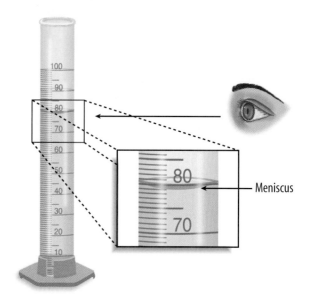

Figure 13 Graduated cylinders measure liquid volume.

Liquid Volume To measure liquids, the unit used is the liter. When a smaller unit is needed, scientists might use a milliliter. Because a milliliter takes up the volume of a cube measuring 1 cm on each side it also can be called a cubic centimeter ($cm^3 = cm \times cm \times cm$).

You can use beakers and graduated cylinders to measure liquid volume. A graduated cylinder, shown in **Figure 13**, is marked from bottom to top in milliliters. In lab, you might use a 10-mL graduated cylinder or a 100-mL graduated cylinder. When measuring liquids, notice that the liquid has a curved surface. Look at the surface at eye level, and measure the bottom of the curve. This is called the meniscus. The graduated cylinder in **Figure 13** contains 79.0 mL, or 79.0 cm^3, of a liquid.

Temperature Scientists often measure temperature using the Celsius scale. Pure water has a freezing point of 0°C and boiling point of 100°C. The unit of measurement is degrees Celsius. Two other scales often used are the Fahrenheit and Kelvin scales.

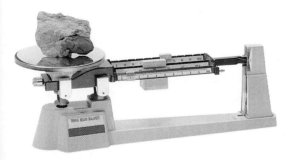

Figure 12 A triple-beam balance is used to determine the mass of an object.

SR-8 • Science Skill Handbook

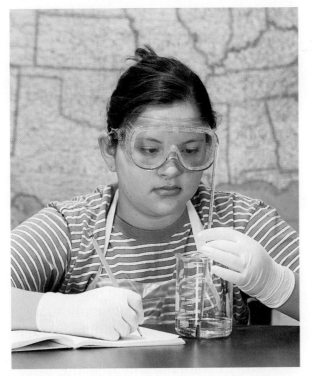

Figure 14 A thermometer measures the temperature of an object.

Scientists use a thermometer to measure temperature. Most thermometers in a laboratory are glass tubes with a bulb at the bottom end containing a liquid such as colored alcohol. The liquid rises or falls with a change in temperature. To read a glass thermometer like the thermometer in **Figure 14,** rotate it slowly until a red line appears. Read the temperature where the red line ends.

Form Operational Definitions An operational definition defines an object by how it functions, works, or behaves. For example, when you are playing hide and seek and a tree is home base, you have created an operational definition for a tree.

Objects can have more than one operational definition. For example, a ruler can be defined as a tool that measures the length of an object (how it is used). It can also be a tool with a series of marks used as a standard when measuring (how it works).

Analyze the Data

To determine the meaning of your observations and investigation results, you will need to look for patterns in the data. Then you must think critically to determine what the data mean. Scientists use several approaches when they analyze the data they have collected and recorded. Each approach is useful for identifying specific patterns.

Interpret Data The word *interpret* means "to explain the meaning of something." When analyzing data from an experiment, try to find out what the data show. Identify the control group and the test group to see whether changes in the independent variable have had an effect. Look for differences in the dependent variable between the control and test groups.

Classify Sorting objects or events into groups based on common features is called classifying. When classifying, first observe the objects or events to be classified. Then select one feature that is shared by some members in the group, but not by all. Place those members that share that feature in a subgroup. You can classify members into smaller and smaller subgroups based on characteristics. Remember that when you classify, you are grouping objects or events for a purpose. Keep your purpose in mind as you select the features to form groups and subgroups.

Compare and Contrast Observations can be analyzed by noting the similarities and differences between two or more objects or events that you observe. When you look at objects or events to see how they are similar, you are comparing them. Contrasting is looking for differences in objects or events.

Science Skill Handbook • **SR-9**

Recognize Cause and Effect A cause is a reason for an action or condition. The effect is that action or condition. When two events happen together, it is not necessarily true that one event caused the other. Scientists must design a controlled investigation to recognize the exact cause and effect.

Draw Conclusions

When scientists have analyzed the data they collected, they proceed to draw conclusions about the data. These conclusions are sometimes stated in words similar to the hypothesis that you formed earlier. They may confirm a hypothesis, or lead you to a new hypothesis.

Infer Scientists often make inferences based on their observations. An inference is an attempt to explain observations or to indicate a cause. An inference is not a fact, but a logical conclusion that needs further investigation. For example, you may infer that a fire has caused smoke. Until you investigate, however, you do not know for sure.

Apply When you draw a conclusion, you must apply those conclusions to determine whether the data supports the hypothesis. If your data do not support your hypothesis, it does not mean that the hypothesis is wrong. It means only that the result of the investigation did not support the hypothesis. Maybe the experiment needs to be redesigned, or some of the initial observations on which the hypothesis was based were incomplete or biased. Perhaps more observation or research is needed to refine your hypothesis. A successful investigation does not always come out the way you originally predicted.

Avoid Bias Sometimes a scientific investigation involves making judgments. When you make a judgment, you form an opinion. It is important to be honest and not to allow any expectations of results to bias your judgments. This is important throughout the entire investigation, from researching to collecting data to drawing conclusions.

Communicate

The communication of ideas is an important part of the work of scientists. A discovery that is not reported will not advance the scientific community's understanding or knowledge. Communication among scientists also is important as a way of improving their investigations.

Scientists communicate in many ways, from writing articles in journals and magazines that explain their investigations and experiments, to announcing important discoveries on television and radio. Scientists also share ideas with colleagues on the Internet or present them as lectures, like the student is doing in **Figure 15.**

Figure 15 A student communicates to his peers about his investigation.

These safety symbols are used in laboratory and field investigations in this book to indicate possible hazards. Learn the meaning of each symbol and refer to this page often. *Remember to wash your hands thoroughly after completing lab procedures.*

PROTECTIVE EQUIPMENT   Do not begin any lab without the proper protection equipment.

 GOGGLES Proper eye protection must be worn when performing or observing science activities that involve items or conditions as listed below.

 APRON Wear an approved apron when using substances that could stain, wet, or destroy cloth.

 SOAP Wash hands with soap and water before removing goggles and after all lab activities.

 GLOVES Wear gloves when working with biological materials, chemicals, animals, or materials that can stain or irritate hands.

LABORATORY HAZARDS

Symbols	Potential Hazards	Precaution	Response
DISPOSAL	contamination of classroom or environment due to improper disposal of materials such as chemicals and live specimens	• DO NOT dispose of hazardous materials in the sink or trash can. • Dispose of wastes as directed by your teacher.	• If hazardous materials are disposed of improperly, notify your teacher immediately.
EXTREME TEMPERATURE	skin burns due to extremely hot or cold materials such as hot glass, liquids, or metals; liquid nitrogen; dry ice	• Use proper protective equipment, such as hot mitts and/or tongs, when handling objects with extreme temperatures.	• If injury occurs, notify your teacher immediately.
SHARP OBJECTS	punctures or cuts from sharp objects such as razor blades, pins, scalpels, and broken glass	• Handle glassware carefully to avoid breakage. • Walk with sharp objects pointed downward, away from you and others.	• If broken glass or injury occurs, notify your teacher immediately.
ELECTRICAL	electric shock or skin burn due to improper grounding, short circuits, liquid spills, or exposed wires	• Check condition of wires and apparatus for fraying or uninsulated wires, and broken or cracked equipment. • Use only GFCI-protected outlets	• DO NOT attempt to fix electrical problems. Notify your teacher immediately.
CHEMICAL	skin irritation or burns, breathing difficulty, and/or poisoning due to touching, swallowing, or inhalation of chemicals such as acids, bases, bleach, metal compounds, iodine, poinsettias, pollen, ammonia, acetone, nail polish remover, heated chemicals, mothballs, and any other chemicals labeled or known to be dangerous	• Wear proper protective equipment such as goggles, apron, and gloves when using chemicals. • Ensure proper room ventilation or use a fume hood when using materials that produce fumes. • NEVER smell fumes directly. • NEVER taste or eat any material in the laboratory.	• If contact occurs, immediately flush affected area with water and notify your teacher. • If a spill occurs, leave the area immediately and notify your teacher.
FLAMMABLE	unexpected fire due to liquids or gases that ignite easily such as rubbing alcohol	• Avoid open flames, sparks, or heat when flammable liquids are present.	• If a fire occurs, leave the area immediately and notify your teacher.
OPEN FLAME	burns or fire due to open flame from matches, Bunsen burners, or burning materials	• Tie back loose hair and clothing. • Keep flame away from all materials. • Follow teacher instructions when lighting and extinguishing flames. • Use proper protection, such as hot mitts or tongs, when handling hot objects.	• If a fire occurs, leave the area immediately and notify your teacher.
ANIMAL SAFETY	injury to or from laboratory animals	• Wear proper protective equipment such as gloves, apron, and goggles when working with animals. • Wash hands after handling animals.	• If injury occurs, notify your teacher immediately.
BIOLOGICAL	infection or adverse reaction due to contact with organisms such as bacteria, fungi, and biological materials such as blood, animal or plant materials	• Wear proper protective equipment such as gloves, goggles, and apron when working with biological materials. • Avoid skin contact with an organism or any part of the organism. • Wash hands after handling organisms.	• If contact occurs, wash the affected area and notify your teacher immediately.
FUME	breathing difficulties from inhalation of fumes from substances such as ammonia, acetone, nail polish remover, heated chemicals, and mothballs	• Wear goggles, apron, and gloves. • Ensure proper room ventilation or use a fume hood when using substances that produce fumes. • NEVER smell fumes directly.	• If a spill occurs, leave area and notify your teacher immediately.
IRRITANT	irritation of skin, mucous membranes, or respiratory tract due to materials such as acids, bases, bleach, pollen, mothballs, steel wool, and potassium permanganate	• Wear goggles, apron, and gloves. • Wear a dust mask to protect against fine particles.	• If skin contact occurs, immediately flush the affected area with water and notify your teacher.
RADIOACTIVE	excessive exposure from alpha, beta, and gamma particles	• Remove gloves and wash hands with soap and water before removing remainder of protective equipment.	• If cracks or holes are found in the container, notify your teacher immediately.

Safety in the Science Laboratory

Introduction to Science Safety

The science laboratory is a safe place to work if you follow standard safety procedures. Being responsible for your own safety helps to make the entire laboratory a safer place for everyone. When performing any lab, read and apply the caution statements and safety symbol listed at the beginning of the lab.

General Safety Rules

1. Complete the *Lab Safety Form* or other safety contract BEFORE starting any science lab.
2. Study the procedure. Ask your teacher any questions. Be sure you understand safety symbols shown on the page.
3. Notify your teacher about allergies or other health conditions that can affect your participation in a lab.
4. Learn and follow use and safety procedures for your equipment. If unsure, ask your teacher.

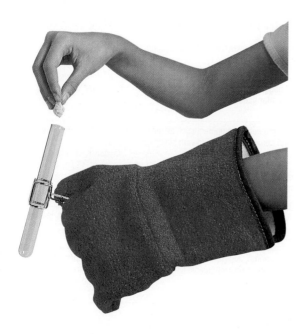

5. Never eat, drink, chew gum, apply cosmetics, or do any personal grooming in the lab. Never use lab glassware as food or drink containers. Keep your hands away from your face and mouth.
6. Know the location and proper use of the safety shower, eye wash, fire blanket, and fire alarm.

Prevent Accidents

1. Use the safety equipment provided to you. Goggles and a safety apron should be worn during investigations.
2. Do NOT use hair spray, mousse, or other flammable hair products. Tie back long hair and tie down loose clothing.
3. Do NOT wear sandals or other open-toed shoes in the lab.
4. Remove jewelry on hands and wrists. Loose jewelry, such as chains and long necklaces, should be removed to prevent them from getting caught in equipment.
5. Do not taste any substances or draw any material into a tube with your mouth.
6. Proper behavior is expected in the lab. Practical jokes and fooling around can lead to accidents and injury.
7. Keep your work area uncluttered.

Laboratory Work

1. Collect and carry all equipment and materials to your work area before beginning a lab.
2. Remain in your own work area unless given permission by your teacher to leave it.

3. Always slant test tubes away from yourself and others when heating them, adding substances to them, or rinsing them.
4. If instructed to smell a substance in a container, hold the container a short distance away and fan vapors toward your nose.
5. Do NOT substitute other chemicals/substances for those in the materials list unless instructed to do so by your teacher.
6. Do NOT take any materials or chemicals outside of the laboratory.
7. Stay out of storage areas unless instructed to be there and supervised by your teacher.

Laboratory Cleanup

1. Turn off all burners, water, and gas, and disconnect all electrical devices.
2. Clean all pieces of equipment and return all materials to their proper places.
3. Dispose of chemicals and other materials as directed by your teacher. Place broken glass and solid substances in the proper containers. Never discard materials in the sink.
4. Clean your work area.
5. Wash your hands with soap and water thoroughly BEFORE removing your goggles.

Emergencies

1. Report any fire, electrical shock, glassware breakage, spill, or injury, no matter how small, to your teacher immediately. Follow his or her instructions.
2. If your clothing should catch fire, STOP, DROP, and ROLL. If possible, smother it with the fire blanket or get under a safety shower. NEVER RUN.
3. If a fire should occur, turn off all gas and leave the room according to established procedures.
4. In most instances, your teacher will clean up spills. Do NOT attempt to clean up spills unless you are given permission and instructions to do so.
5. If chemicals come into contact with your eyes or skin, notify your teacher immediately. Use the eyewash, or flush your skin or eyes with large quantities of water.
6. The fire extinguisher and first-aid kit should only be used by your teacher unless it is an extreme emergency and you have been given permission.
7. If someone is injured or becomes ill, only a professional medical provider or someone certified in first aid should perform first-aid procedures.

Math Skill Handbook

Math Review

Use Fractions

A fraction compares a part to a whole. In the fraction $\frac{2}{3}$, the 2 represents the part and is the numerator. The 3 represents the whole and is the denominator.

Reduce Fractions To reduce a fraction, you must find the largest factor that is common to both the numerator and the denominator, the greatest common factor (GCF). Divide both numbers by the GCF. The fraction has then been reduced, or it is in its simplest form.

Example

Twelve of the 20 chemicals in the science lab are in powder form. What fraction of the chemicals used in the lab are in powder form?

Step 1 Write the fraction.

$$\frac{\text{part}}{\text{whole}} = \frac{12}{20}$$

Step 2 To find the GCF of the numerator and denominator, list all of the factors of each number.

Factors of 12: 1, 2, 3, 4, 6, 12 (the numbers that divide evenly into 12)

Factors of 20: 1, 2, 4, 5, 10, 20 (the numbers that divide evenly into 20)

Step 3 List the common factors.

1, 2, 4

Step 4 Choose the greatest factor in the list. The GCF of 12 and 20 is 4.

Step 5 Divide the numerator and denominator by the GCF.

$$\frac{12 \div 4}{20 \div 4} = \frac{3}{5}$$

In the lab, $\frac{3}{5}$ of the chemicals are in powder form.

Practice Problem At an amusement park, 66 of 90 rides have a height restriction. What fraction of the rides, in its simplest form, has a height restriction?

Add and Subtract Fractions with Like Denominators To add or subtract fractions with the same denominator, add or subtract the numerators and write the sum or difference over the denominator. After finding the sum or difference, find the simplest form for your fraction.

Example 1

In the forest outside your house, $\frac{1}{8}$ of the animals are rabbits, $\frac{3}{8}$ are squirrels, and the remainder are birds and insects. How many are mammals?

Step 1 Add the numerators.

$$\frac{1}{8} + \frac{3}{8} = \frac{(1+3)}{8} = \frac{4}{8}$$

Step 2 Find the GCF.

$\frac{4}{8}$ (GCF, 4)

Step 3 Divide the numerator and denominator by the GCF.

$$\frac{4 \div 4}{8 \div 4} = \frac{1}{2}$$

$\frac{1}{2}$ of the animals are mammals.

Example 2

If $\frac{7}{16}$ of the Earth is covered by freshwater, and $\frac{1}{16}$ of that is in glaciers, how much freshwater is not frozen?

Step 1 Subtract the numerators.

$$\frac{7}{16} - \frac{1}{16} = \frac{(7-1)}{16} = \frac{6}{16}$$

Step 2 Find the GCF.

$\frac{6}{16}$ (GCF, 2)

Step 3 Divide the numerator and denominator by the GCF.

$$\frac{6 \div 2}{16 \div 2} = \frac{3}{8}$$

$\frac{3}{8}$ of the freshwater is not frozen.

Practice Problem A bicycle rider is riding at a rate of 15 km/h for $\frac{4}{9}$ of his ride, 10 km/h for $\frac{2}{9}$ of his ride, and 8 km/h for the remainder of the ride. How much of his ride is he riding at a rate greater than 8 km/h?

Add and Subtract Fractions with Unlike Denominators To add or subtract fractions with unlike denominators, first find the least common denominator (LCD). This is the smallest number that is a common multiple of both denominators. Rename each fraction with the LCD, and then add or subtract. Find the simplest form if necessary.

Example 1

A chemist makes a paste that is $\frac{1}{2}$ table salt (NaCl), $\frac{1}{3}$ sugar ($C_6H_{12}O_6$), and the remainder is water (H_2O). How much of the paste is a solid?

Step 1 Find the LCD of the fractions.

$\frac{1}{2} + \frac{1}{3}$ (LCD, 6)

Step 2 Rename each numerator and each denominator with the LCD.

Step 3 Add the numerators.

$\frac{3}{6} + \frac{2}{6} = \frac{(3+2)}{6} = \frac{5}{6}$

$\frac{5}{6}$ of the paste is a solid.

Example 2

The average precipitation in Grand Junction, CO, is $\frac{7}{10}$ inch in November, and $\frac{3}{5}$ inch in December. What is the total average precipitation?

Step 1 Find the LCD of the fractions.

$\frac{7}{10} + \frac{3}{5}$ (LCD, 10)

Step 2 Rename each numerator and each denominator with the LCD.

Step 3 Add the numerators.

$\frac{7}{10} + \frac{6}{10} = \frac{(7+6)}{10} = \frac{13}{10}$

$\frac{13}{10}$ inches total precipitation, or $1\frac{3}{10}$ inches.

Practice Problem On an electric bill, about $\frac{1}{8}$ of the energy is from solar energy and about $\frac{1}{10}$ is from wind power. How much of the total bill is from solar energy and wind power combined?

Example 3

In your body, $\frac{7}{10}$ of your muscle contractions are involuntary (cardiac and smooth muscle tissue). Smooth muscle makes $\frac{3}{15}$ of your muscle contractions. How many of your muscle contractions are made by cardiac muscle?

Step 1 Find the LCD of the fractions.

$\frac{7}{10} - \frac{3}{15}$ (LCD, 30)

Step 2 Rename each numerator and each denominator with the LCD.

$\frac{7 \times 3}{10 \times 3} = \frac{21}{30}$

$\frac{3 \times 2}{15 \times 2} = \frac{6}{30}$

Step 3 Subtract the numerators.

$\frac{21}{30} - \frac{6}{30} = \frac{(21-6)}{30} = \frac{15}{30}$

Step 4 Find the GCF.

$\frac{15}{30}$ (GCF, 15)

$\frac{1}{2}$

$\frac{1}{2}$ of all muscle contractions are cardiac muscle.

Example 4

Tony wants to make cookies that call for $\frac{3}{4}$ of a cup of flour, but he only has $\frac{1}{3}$ of a cup. How much more flour does he need?

Step 1 Find the LCD of the fractions.

$\frac{3}{4} - \frac{1}{3}$ (LCD, 12)

Step 2 Rename each numerator and each denominator with the LCD.

$\frac{3 \times 3}{4 \times 3} = \frac{9}{12}$

$\frac{1 \times 4}{3 \times 4} = \frac{4}{12}$

Step 3 Subtract the numerators.

$\frac{9}{12} - \frac{4}{12} = \frac{(9-4)}{12} = \frac{5}{12}$

$\frac{5}{12}$ of a cup of flour

Practice Problem Using the information provided to you in Example 3 above, determine how many muscle contractions are voluntary (skeletal muscle).

Math Skill Handbook • **SR-15**

Multiply Fractions To multiply with fractions, multiply the numerators and multiply the denominators. Find the simplest form if necessary.

> **Example**
>
> Multiply $\frac{3}{5}$ by $\frac{1}{3}$.
>
> **Step 1** Multiply the numerators and denominators.
>
> $\frac{3}{5} \times \frac{1}{3} = \frac{(3 \times 1)}{(5 \times 3)} \quad \frac{3}{15}$
>
> **Step 2** Find the GCF.
>
> $\frac{3}{15}$ (GCF, 3)
>
> **Step 3** Divide the numerator and denominator by the GCF.
>
> $\frac{3 \div 3}{15 \div 3} = \frac{1}{5}$
>
> $\frac{3}{5}$ multiplied by $\frac{1}{3}$ is $\frac{1}{5}$.

Practice Problem Multiply $\frac{3}{14}$ by $\frac{5}{16}$.

Find a Reciprocal Two numbers whose product is 1 are called multiplicative inverses, or reciprocals.

> **Example**
>
> Find the reciprocal of $\frac{3}{8}$.
>
> **Step 1** Inverse the fraction by putting the denominator on top and the numerator on the bottom.
>
> $\frac{8}{3}$
>
> The reciprocal of $\frac{3}{8}$ is $\frac{8}{3}$.

Practice Problem Find the reciprocal of $\frac{4}{9}$.

Divide Fractions To divide one fraction by another fraction, multiply the dividend by the reciprocal of the divisor. Find the simplest form if necessary.

> **Example 1**
>
> Divide $\frac{1}{9}$ by $\frac{1}{3}$.
>
> **Step 1** Find the reciprocal of the divisor.
>
> The reciprocal of $\frac{1}{3}$ is $\frac{3}{1}$.
>
> **Step 2** Multiply the dividend by the reciprocal of the divisor.
>
> $\frac{\frac{1}{9}}{\frac{1}{3}} = \frac{1}{9} \times \frac{3}{1} = \frac{(1 \times 3)}{(9 \times 1)} = \frac{3}{9}$
>
> **Step 3** Find the GCF.
>
> $\frac{3}{9}$ (GCF, 3)
>
> **Step 4** Divide the numerator and denominator by the GCF.
>
> $\frac{3 \div 3}{9 \div 3} = \frac{1}{3}$
>
> $\frac{1}{9}$ divided by $\frac{1}{3}$ is $\frac{1}{3}$.

> **Example 2**
>
> Divide $\frac{3}{5}$ by $\frac{1}{4}$.
>
> **Step 1** Find the reciprocal of the divisor.
>
> The reciprocal of $\frac{1}{4}$ is $\frac{4}{1}$.
>
> **Step 2** Multiply the dividend by the reciprocal of the divisor.
>
> $\frac{\frac{3}{5}}{\frac{1}{4}} = \frac{3}{5} \times \frac{4}{1} = \frac{(3 \times 4)}{(5 \times 1)} = \frac{12}{5}$
>
> $\frac{3}{5}$ divided by $\frac{1}{4}$ is $\frac{12}{5}$ or $2\frac{2}{5}$.

Practice Problem Divide $\frac{3}{11}$ by $\frac{7}{10}$.

Use Ratios

When you compare two numbers by division, you are using a ratio. Ratios can be written 3 to 5, 3:5, or $\frac{3}{5}$. Ratios, like fractions, also can be written in simplest form.

Ratios can represent one type of probability, called odds. This is a ratio that compares the number of ways a certain outcome occurs to the number of possible outcomes. For example, if you flip a coin 100 times, what are the odds that it will come up heads? There are two possible outcomes, heads or tails, so the odds of coming up heads are 50:100. Another way to say this is that 50 out of 100 times the coin will come up heads. In its simplest form, the ratio is 1:2.

Example 1

A chemical solution contains 40 g of salt and 64 g of baking soda. What is the ratio of salt to baking soda as a fraction in simplest form?

Step 1 Write the ratio as a fraction.
$$\frac{salt}{baking\ soda} = \frac{40}{64}$$

Step 2 Express the fraction in simplest form. The GCF of 40 and 64 is 8.
$$\frac{40}{64} = \frac{40 \div 8}{64 \div 8} = \frac{5}{8}$$

The ratio of salt to baking soda in the sample is 5:8.

Example 2

Sean rolls a 6-sided die 6 times. What are the odds that the side with a 3 will show?

Step 1 Write the ratio as a fraction.
$$\frac{number\ of\ sides\ with\ a\ 3}{number\ of\ possible\ sides} = \frac{1}{6}$$

Step 2 Multiply by the number of attempts.
$$\frac{1}{6} \times 6\ attempts = \frac{6}{6}\ attempts = 1\ attempt$$

1 attempt out of 6 will show a 3.

Practice Problem Two metal rods measure 100 cm and 144 cm in length. What is the ratio of their lengths in simplest form?

Use Decimals

A fraction with a denominator that is a power of ten can be written as a decimal. For example, 0.27 means $\frac{27}{100}$. The decimal point separates the ones place from the tenths place.

Any fraction can be written as a decimal using division. For example, the fraction $\frac{5}{8}$ can be written as a decimal by dividing 5 by 8. Written as a decimal, it is 0.625.

Add or Subtract Decimals When adding and subtracting decimals, line up the decimal points before carrying out the operation.

Example 1

Find the sum of 47.68 and 7.80.

Step 1 Line up the decimal places when you write the numbers.

```
  47.68
+  7.80
```

Step 2 Add the decimals.
```
  ¹¹
  47.68
+  7.80
  55.48
```

The sum of 47.68 and 7.80 is 55.48.

Example 2

Find the difference of 42.17 and 15.85.

Step 1 Line up the decimal places when you write the number.

```
  42.17
- 15.85
```

Step 2 Subtract the decimals.
```
  ³¹¹
  42.17
- 15.85
  26.32
```

The difference of 42.17 and 15.85 is 26.32.

Practice Problem Find the sum of 1.245 and 3.842.

Multiply Decimals To multiply decimals, multiply the numbers like numbers without decimal points. Count the decimal places in each factor. The product will have the same number of decimal places as the sum of the decimal places in the factors.

Example

Multiply 2.4 by 5.9.

Step 1 Multiply the factors like two whole numbers.

$24 \times 59 = 1416$

Step 2 Find the sum of the number of decimal places in the factors. Each factor has one decimal place, for a sum of two decimal places.

Step 3 The product will have two decimal places.

14.16

The product of 2.4 and 5.9 is 14.16.

Practice Problem Multiply 4.6 by 2.2.

Divide Decimals When dividing decimals, change the divisor to a whole number. To do this, multiply both the divisor and the dividend by the same power of ten. Then place the decimal point in the quotient directly above the decimal point in the dividend. Then divide as you do with whole numbers.

Example

Divide 8.84 by 3.4.

Step 1 Multiply both factors by 10.

$3.4 \times 10 = 34, 8.84 \times 10 = 88.4$

Step 2 Divide 88.4 by 34.

```
      2.6
   34)88.4
     -68
      204
     -204
        0
```

8.84 divided by 3.4 is 2.6.

Practice Problem Divide 75.6 by 3.6.

Use Proportions

An equation that shows that two ratios are equivalent is a proportion. The ratios $\frac{2}{4}$ and $\frac{5}{10}$ are equivalent, so they can be written as $\frac{2}{4} = \frac{5}{10}$. This equation is a proportion.

When two ratios form a proportion, the cross products are equal. To find the cross products in the proportion $\frac{2}{4} = \frac{5}{10}$, multiply the 2 and the 10, and the 4 and the 5. Therefore $2 \times 10 = 4 \times 5$, or $20 = 20$.

Because you know that both ratios are equal, you can use cross products to find a missing term in a proportion. This is known as solving the proportion.

Example

The heights of a tree and a pole are proportional to the lengths of their shadows. The tree casts a shadow of 24 m when a 6-m pole casts a shadow of 4 m. What is the height of the tree?

Step 1 Write a proportion.

$$\frac{\text{height of tree}}{\text{height of pole}} = \frac{\text{length of tree's shadow}}{\text{length of pole's shadow}}$$

Step 2 Substitute the known values into the proportion. Let h represent the unknown value, the height of the tree.

$$\frac{h}{6} \times \frac{24}{4}$$

Step 3 Find the cross products.

$h \times 4 = 6 \times 24$

Step 4 Simplify the equation.

$4h \times 144$

Step 5 Divide each side by 4.

$$\frac{4h}{4} \times \frac{144}{4}$$

$h = 36$

The height of the tree is 36 m.

Practice Problem The ratios of the weights of two objects on the Moon and on Earth are in proportion. A rock weighing 3 N on the Moon weighs 18 N on Earth. How much would a rock that weighs 5 N on the Moon weigh on Earth?

Use Percentages

The word *percent* means "out of one hundred." It is a ratio that compares a number to 100. Suppose you read that 77 percent of Earth's surface is covered by water. That is the same as reading that the fraction of Earth's surface covered by water is $\frac{77}{100}$. To express a fraction as a percent, first find the equivalent decimal for the fraction. Then, multiply the decimal by 100 and add the percent symbol.

Example 1

Express $\frac{13}{20}$ as a percent.

Step 1 Find the equivalent decimal for the fraction.

$$\begin{array}{r} 0.65 \\ 20\overline{)13.00} \\ \underline{12\ 0} \\ 1\ 00 \\ \underline{1\ 00} \\ 0 \end{array}$$

Step 2 Rewrite the fraction $\frac{13}{20}$ as 0.65.

Step 3 Multiply 0.65 by 100 and add the % symbol.

$$0.65 \times 100 = 65 = 65\%$$

So, $\frac{13}{20} = 65\%$.

This also can be solved as a proportion.

Example 2

Express $\frac{13}{20}$ as a percent.

Step 1 Write a proportion.

$$\frac{13}{20} = \frac{x}{100}$$

Step 2 Find the cross products.

$$1300 = 20x$$

Step 3 Divide each side by 20.

$$\frac{1300}{20} = \frac{20x}{20}$$
$$65\% = x$$

Practice Problem In one year, 73 of 365 days were rainy in one city. What percent of the days in that city were rainy?

Solve One-Step Equations

A statement that two expressions are equal is an equation. For example, $A = B$ is an equation that states that A is equal to B.

An equation is solved when a variable is replaced with a value that makes both sides of the equation equal. To make both sides equal the inverse operation is used. Addition and subtraction are inverses, and multiplication and division are inverses.

Example 1

Solve the equation $x - 10 = 35$.

Step 1 Find the solution by adding 10 to each side of the equation.

$$x - 10 = 35$$
$$x - 10 + 10 = 35 - 10$$
$$x = 45$$

Step 2 Check the solution.

$$x - 10 = 35$$
$$45 - 10 = 35$$
$$35 = 35$$

Both sides of the equation are equal, so $x = 45$.

Example 2

In the formula $a = bc$, find the value of c if $a = 20$ and $b = 2$.

Step 1 Rearrange the formula so the unknown value is by itself on one side of the equation by dividing both sides by b.

$$a = bc$$
$$\frac{a}{b} = \frac{bc}{b}$$
$$\frac{a}{b} = c$$

Step 2 Replace the variables a and b with the values that are given.

$$\frac{a}{b} = c$$
$$\frac{20}{2} = c$$
$$10 = c$$

Step 3 Check the solution.

$$a = bc$$
$$20 = 2 \times 10$$
$$20 = 20$$

Both sides of the equation are equal, so $c = 10$ is the solution when $a = 20$ and $b = 2$.

Practice Problem In the formula $h = gd$, find the value of d if $g = 12.3$ and $h = 17.4$.

Use Statistics

The branch of mathematics that deals with collecting, analyzing, and presenting data is statistics. In statistics, there are three common ways to summarize data with a single number—the mean, the median, and the mode.

The **mean** of a set of data is the arithmetic average. It is found by adding the numbers in the data set and dividing by the number of items in the set.

The **median** is the middle number in a set of data when the data are arranged in numerical order. If there were an even number of data points, the median would be the mean of the two middle numbers.

The **mode** of a set of data is the number or item that appears most often.

Another number that often is used to describe a set of data is the range. The **range** is the difference between the largest number and the smallest number in a set of data.

Example

The speeds (in m/s) for a race car during five different time trials are 39, 37, 44, 36, and 44.

To find the mean:

Step 1 Find the sum of the numbers.

39 + 37 + 44 + 36 + 44 = 200

Step 2 Divide the sum by the number of items, which is 5.

200 ÷ 5 = 40

The mean is 40 m/s.

To find the median:

Step 1 Arrange the measures from least to greatest.

36, 37, 39, 44, 44

Step 2 Determine the middle measure.

36, 37, <u>39</u>, 44, 44

The median is 39 m/s.

To find the mode:

Step 1 Group the numbers that are the same together.

44, 44, 36, 37, 39

Step 2 Determine the number that occurs most in the set.

<u>44, 44,</u> 36, 37, 39

The mode is 44 m/s.

To find the range:

Step 1 Arrange the measures from greatest to least.

44, 44, 39, 37, 36

Step 2 Determine the greatest and least measures in the set.

<u>44,</u> 44, 39, 37, <u>36</u>

Step 3 Find the difference between the greatest and least measures.

44 − 36 = 8

The range is 8 m/s.

Practice Problem Find the mean, median, mode, and range for the data set 8, 4, 12, 8, 11, 14, 16.

A **frequency table** shows how many times each piece of data occurs, usually in a survey. **Table 1** below shows the results of a student survey on favorite color.

Table 1 Student Color Choice		
Color	Tally	Frequency
red	IIII	4
blue	IIII	5
black	II	2
green	III	3
purple	IIII II	7
yellow	IIII I	6

Based on the frequency table data, which color is the favorite?

Use Geometry

The branch of mathematics that deals with the measurement, properties, and relationships of points, lines, angles, surfaces, and solids is called geometry.

Perimeter The **perimeter** (P) is the distance around a geometric figure. To find the perimeter of a rectangle, add the length and width and multiply that sum by two, or $2(l + w)$. To find perimeters of irregular figures, add the length of the sides.

Example 1

Find the perimeter of a rectangle that is 3 m long and 5 m wide.

Step 1 You know that the perimeter is 2 times the sum of the width and length.

$P = 2(3\text{ m} + 5\text{ m})$

Step 2 Find the sum of the width and length.

$P = 2(8\text{ m})$

Step 3 Multiply by 2.

$P = 16\text{ m}$

The perimeter is 16 m.

Example 2

Find the perimeter of a shape with sides measuring 2 cm, 5 cm, 6 cm, 3 cm.

Step 1 You know that the perimeter is the sum of all the sides.

$P = 2 + 5 + 6 + 3$

Step 2 Find the sum of the sides.

$P = 2 + 5 + 6 + 3$

$P = 16$

The perimeter is 16 cm.

Practice Problem Find the perimeter of a rectangle with a length of 18 m and a width of 7 m.

Practice Problem Find the perimeter of a triangle measuring 1.6 cm by 2.4 cm by 2.4 cm.

Area of a Rectangle The **area** (A) is the number of square units needed to cover a surface. To find the area of a rectangle, multiply the length times the width, or $l \times w$. When finding area, the units also are multiplied. Area is given in square units.

Example

Find the area of a rectangle with a length of 1 cm and a width of 10 cm.

Step 1 You know that the area is the length multiplied by the width.

$A = (1\text{ cm} \times 10\text{ cm})$

Step 2 Multiply the length by the width. Also multiply the units.

$A = 10\text{ cm}^2$

The area is 10 cm^2.

Practice Problem Find the area of a square whose sides measure 4 m.

Area of a Triangle To find the area of a triangle, use the formula:

$A = \frac{1}{2}(\text{base} \times \text{height})$

The base of a triangle can be any of its sides. The height is the perpendicular distance from a base to the opposite endpoint, or vertex.

Example

Find the area of a triangle with a base of 18 m and a height of 7 m.

Step 1 You know that the area is $\frac{1}{2}$ the base times the height.

$A = \frac{1}{2}(18\text{ m} \times 7\text{ m})$

Step 2 Multiply $\frac{1}{2}$ by the product of 18×7. Multiply the units.

$A = \frac{1}{2}(126\text{ m}^2)$

$A = 63\text{ m}^2$

The area is 63 m^2.

Practice Problem Find the area of a triangle with a base of 27 cm and a height of 17 cm.

Circumference of a Circle The **diameter** (*d*) of a circle is the distance across the circle through its center, and the **radius** (r) is the distance from the center to any point on the circle. The radius is half of the diameter. The distance around the circle is called the **circumference** (C). The formula for finding the circumference is:

$C = 2\pi r$ or $C = \pi d$

The circumference divided by the diameter is always equal to 3.1415926... This nonterminating and nonrepeating number is represented by the Greek letter π (pi). An approximation often used for π is 3.14.

Example 1

Find the circumference of a circle with a radius of 3 m.

Step 1 You know the formula for the circumference is 2 times the radius times π.

$C = 2\pi(3)$

Step 2 Multiply 2 times the radius.

$C = 6\pi$

Step 3 Multiply by π.

$C \approx 19$ m

The circumference is about 19 m.

Example 2

Find the circumference of a circle with a diameter of 24.0 cm.

Step 1 You know the formula for the circumference is the diameter times π.

$C = \pi(24.0)$

Step 2 Multiply the diameter by π.

$C \approx 75.4$ cm

The circumference is about 75.4 cm.

Practice Problem Find the circumference of a circle with a radius of 19 cm.

Area of a Circle The formula for the area of a circle is: $A = \pi r^2$

Example 1

Find the area of a circle with a radius of 4.0 cm.

Step 1 $A = \pi(4.0)^2$

Step 2 Find the square of the radius.

$A = 16\pi$

Step 3 Multiply the square of the radius by π.

$A \approx 50$ cm^2

The area of the circle is about 50 cm^2.

Example 2

Find the area of a circle with a radius of 225 m.

Step 1 $A = \pi(225)^2$

Step 2 Find the square of the radius.

$A = 50625\pi$

Step 3 Multiply the square of the radius by π.

$A \approx 159043.1$

The area of the circle is about 159043.1 m^2.

Example 3

Find the area of a circle whose diameter is 20.0 mm.

Step 1 Remember that the radius is half of the diameter.

$A = \pi\left(\dfrac{20.0}{2}\right)^2$

Step 2 Find the radius.

$A = \pi(10.0)^2$

Step 3 Find the square of the radius.

$A = 100\pi$

Step 4 Multiply the square of the radius by π.

$A \approx 314$ mm^2

The area of the circle is about 314 mm^2.

Practice Problem Find the area of a circle with a radius of 16 m.

Volume The measure of space occupied by a solid is the **volume** (V). To find the volume of a rectangular solid multiply the length times width times height, or $V = l \times w \times h$. It is measured in cubic units, such as cubic centimeters (cm^3).

Example

Find the volume of a rectangular solid with a length of 2.0 m, a width of 4.0 m, and a height of 3.0 m.

Step 1 You know the formula for volume is the length times the width times the height.

$V = 2.0 \text{ m} \times 4.0 \text{ m} \times 3.0 \text{ m}$

Step 2 Multiply the length times the width times the height.

$V = 24 \text{ m}^3$

The volume is 24 m^3.

Practice Problem Find the volume of a rectangular solid that is 8 m long, 4 m wide, and 4 m high.

To find the volume of other solids, multiply the area of the base times the height.

Example 1

Find the volume of a solid that has a triangular base with a length of 8.0 m and a height of 7.0 m. The height of the entire solid is 15.0 m.

Step 1 You know that the base is a triangle, and the area of a triangle is $\frac{1}{2}$ the base times the height, and the volume is the area of the base times the height.

$V = \left[\frac{1}{2}(b \times h)\right] \times 15$

Step 2 Find the area of the base.

$V = \left[\frac{1}{2}(8 \times 7)\right] \times 15$

$V = \left(\frac{1}{2} \times 56\right) \times 15$

Step 3 Multiply the area of the base by the height of the solid.

$V = 28 \times 15$

$V = 420 \text{ m}^3$

The volume is 420 m^3.

Example 2

Find the volume of a cylinder that has a base with a radius of 12.0 cm, and a height of 21.0 cm.

Step 1 You know that the base is a circle, and the area of a circle is the square of the radius times π, and the volume is the area of the base times the height.

$V = (\pi r^2) \times 21$

$V = (\pi 12^2) \times 21$

Step 2 Find the area of the base.

$V = 144\pi \times 21$

$V = 452 \times 21$

Step 3 Multiply the area of the base by the height of the solid.

$V \approx 9{,}500 \text{ cm}^3$

The volume is about 9,500 cm^3.

Example 3

Find the volume of a cylinder that has a diameter of 15 mm and a height of 4.8 mm.

Step 1 You know that the base is a circle with an area equal to the square of the radius times π. The radius is one-half the diameter. The volume is the area of the base times the height.

$V = (\pi r^2) \times 4.8$

$V = \left[\pi\left(\frac{1}{2} \times 15\right)^2\right] \times 4.8$

$V = (\pi 7.5^2) \times 4.8$

Step 2 Find the area of the base.

$V = 56.25\pi \times 4.8$

$V \approx 176.71 \times 4.8$

Step 3 Multiply the area of the base by the height of the solid.

$V \approx 848.2$

The volume is about 848.2 mm^3.

Practice Problem Find the volume of a cylinder with a diameter of 7 cm in the base and a height of 16 cm.

Math Skill Handbook • **SR-23**

Science Applications

Measure in SI

The metric system of measurement was developed in 1795. A modern form of the metric system, called the International System (SI), was adopted in 1960 and provides the standard measurements that all scientists around the world can understand.

The SI system is convenient because unit sizes vary by powers of 10. Prefixes are used to name units. Look at **Table 2** for some common SI prefixes and their meanings.

Table 2 Common SI Prefixes

Prefix	Symbol	Meaning	
kilo–	k	1,000	thousandth
hecto–	h	100	hundred
deka–	da	10	ten
deci–	d	0.1	tenth
centi–	c	0.01	hundreth
milli–	m	0.001	thousandth

Example

How many grams equal one kilogram?

Step 1 Find the prefix *kilo–* in **Table 2**.

Step 2 Using **Table 2**, determine the meaning of *kilo–*. According to the table, it means 1,000. When the prefix *kilo–* is added to a unit, it means that there are 1,000 of the units in a "kilounit."

Step 3 Apply the prefix to the units in the question. The units in the question are grams. There are 1,000 grams in a kilogram.

Practice Problem Is a milligram larger or smaller than a gram? How many of the smaller units equal one larger unit? What fraction of the larger unit does one smaller unit represent?

Dimensional Analysis

Convert SI Units In science, quantities such as length, mass, and time sometimes are measured using different units. A process called dimensional analysis can be used to change one unit of measure to another. This process involves multiplying your starting quantity and units by one or more conversion factors. A conversion factor is a ratio equal to one and can be made from any two equal quantities with different units. If 1,000 mL equal 1 L then two ratios can be made.

$$\frac{1{,}000 \text{ mL}}{1 \text{ L}} = \frac{1 \text{ L}}{1{,}000 \text{ mL}} = 1$$

One can convert between units in the SI system by using the equivalents in **Table 2** to make conversion factors.

Example

How many cm are in 4 m?

Step 1 Write conversion factors for the units given. From **Table 2**, you know that 100 cm = 1 m. The conversion factors are

$$\frac{100 \text{ cm}}{1 \text{ m}} \text{ and } \frac{1 \text{ m}}{100 \text{ cm}}$$

Step 2 Decide which conversion factor to use. Select the factor that has the units you are converting from (m) in the denominator and the units you are converting to (cm) in the numerator.

$$\frac{100 \text{ cm}}{1 \text{ m}}$$

Step 3 Multiply the starting quantity and units by the conversion factor. Cancel the starting units with the units in the denominator. There are 400 cm in 4 m.

$$4 \text{ m} \times \frac{100 \text{ cm}}{1 \text{ m}} = 400 \text{ cm}$$

Practice Problem How many milligrams are in one kilogram? (Hint: You will need to use two conversion factors from **Table 2**.)

Table 3 Unit System Equivalents

Type of Measurement	Equivalent
Length	1 in = 2.54 cm 1 yd = 0.91 m 1 mi = 1.61 km
Mass and weight*	1 oz = 28.35 g 1 lb = 0.45 kg 1 ton (short) = 0.91 tonnes (metric tons) 1 lb = 4.45 N
Volume	1 in^3 = 16.39 cm^3 1 qt = 0.95 L 1 gal = 3.78 L
Area	1 in^2 = 6.45 cm^2 1 yd^2 = 0.83 m^2 1 mi^2 = 2.59 km^2 1 acre = 0.40 hectares
Temperature	°C = $\frac{(°F - 32)}{1.8}$ K = °C + 273

*Weight is measured in standard Earth gravity.

Convert Between Unit Systems Table 3 gives a list of equivalents that can be used to convert between English and SI units.

Example

If a meterstick has a length of 100 cm, how long is the meterstick in inches?

Step 1 Write the conversion factors for the units given. From **Table 3,** 1 in = 2.54 cm.

$$\frac{1 \text{ in}}{2.54 \text{ cm}} \text{ and } \frac{2.54 \text{ cm}}{1 \text{ in}}$$

Step 2 Determine which conversion factor to use. You are converting from cm to in. Use the conversion factor with cm on the bottom.

$$\frac{1 \text{ in}}{2.54 \text{ cm}}$$

Step 3 Multiply the starting quantity and units by the conversion factor. Cancel the starting units with the units in the denominator. Round your answer to the nearest tenth.

$$100 \text{ cm} \times \frac{1 \text{ in}}{2.54 \text{ cm}} = 39.37 \text{ in}$$

The meterstick is about 39.4 in long.

Practice Problem 1 A book has a mass of 5 lb. What is the mass of the book in kg?

Practice Problem 2 Use the equivalent for in and cm (1 in = 2.54 cm) to show how 1 in^3 ≈ 16.39 cm^3.

Math Skill Handbook • **SR-25**

Precision and Significant Digits

When you make a measurement, the value you record depends on the precision of the measuring instrument. This precision is represented by the number of significant digits recorded in the measurement. When counting the number of significant digits, all digits are counted except zeros at the end of a number with no decimal point such as 2,050, and zeros at the beginning of a decimal such as 0.03020. When adding or subtracting numbers with different precision, round the answer to the smallest number of decimal places of any number in the sum or difference. When multiplying or dividing, the answer is rounded to the smallest number of significant digits of any number being multiplied or divided.

Example

The lengths 5.28 and 5.2 are measured in meters. Find the sum of these lengths and record your answer using the correct number of significant digits.

Step 1 Find the sum.

    5.28 m   2 digits after the decimal
+ 5.2 m   1 digit after the decimal
   10.48 m

Step 2 Round to one digit after the decimal because the least number of digits after the decimal of the numbers being added is 1.

The sum is 10.5 m.

Practice Problem 1 How many significant digits are in the measurement 7,071,301 m? How many significant digits are in the measurement 0.003010 g?

Practice Problem 2 Multiply 5.28 and 5.2 using the rule for multiplying and dividing. Record the answer using the correct number of significant digits.

Scientific Notation

Many times numbers used in science are very small or very large. Because these numbers are difficult to work with scientists use scientific notation. To write numbers in scientific notation, move the decimal point until only one non-zero digit remains on the left. Then count the number of places you moved the decimal point and use that number as a power of ten. For example, the average distance from the Sun to Mars is 227,800,000,000 m. In scientific notation, this distance is 2.278×10^{11} m. Because you moved the decimal point to the left, the number is a positive power of ten.

The mass of an electron is about 0.000 000 000 000 000 000 000 000 000 000 911 kg. Expressed in scientific notation, this mass is 9.11×10^{-31} kg. Because the decimal point was moved to the right, the number is a negative power of ten.

Example

Earth is 149,600,000 km from the Sun. Express this in scientific notation.

Step 1 Move the decimal point until one non-zero digit remains on the left.

1.496 000 00

Step 2 Count the number of decimal places you have moved. In this case, eight.

Step 2 Show that number as a power of ten, 10^8.

Earth is 1.496×10^8 km from the Sun.

Practice Problem 1 How many significant digits are in 149,600,000 km? How many significant digits are in 1.496×10^8 km?

Practice Problem 2 Parts used in a high performance car must be measured to 7×10^{-6} m. Express this number as a decimal.

Practice Problem 3 A CD is spinning at 539 revolutions per minute. Express this number in scientific notation.

Make and Use Graphs

Data in tables can be displayed in a graph—a visual representation of data. Common graph types include line graphs, bar graphs, and circle graphs.

Line Graph A line graph shows a relationship between two variables that change continuously. The independent variable is changed and is plotted on the x-axis. The dependent variable is observed, and is plotted on the y-axis.

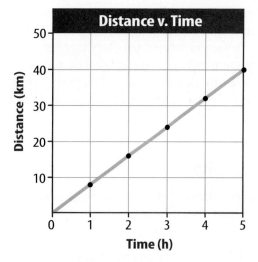

Figure 8 This line graph shows the relationship between distance and time during a bicycle ride.

Example

Draw a line graph of the data below from a cyclist in a long-distance race.

Table 4 Bicycle Race Data

Time (h)	Distance (km)
0	0
1	8
2	16
3	24
4	32
5	40

Step 1 Determine the x-axis and y-axis variables. Time varies independently of distance and is plotted on the x-axis. Distance is dependent on time and is plotted on the y-axis.

Step 2 Determine the scale of each axis. The x-axis data ranges from 0 to 5. The y-axis data ranges from 0 to 50.

Step 3 Using graph paper, draw and label the axes. Include units in the labels.

Step 4 Draw a point at the intersection of the time value on the x-axis and corresponding distance value on the y-axis. Connect the points and label the graph with a title, as shown in **Figure 8**.

Practice Problem A puppy's shoulder height is measured during the first year of her life. The following measurements were collected: (3 mo, 52 cm), (6 mo, 72 cm), (9 mo, 83 cm), (12 mo, 86 cm). Graph this data.

Find a Slope The slope of a straight line is the ratio of the vertical change, rise, to the horizontal change, run.

$$\text{Slope} = \frac{\text{vertical change (rise)}}{\text{horizontal change (run)}} = \frac{\text{change in } y}{\text{change in } x}$$

Example

Find the slope of the graph in **Figure 8**.

Step 1 You know that the slope is the change in y divided by the change in x.

$$\text{Slope} = \frac{\text{change in } y}{\text{change in } x}$$

Step 2 Determine the data points you will be using. For a straight line, choose the two sets of points that are the farthest apart.

$$\text{Slope} = \frac{(40 - 0) \text{ km}}{(5 - 0) \text{ h}}$$

Step 3 Find the change in y and x.

$$\text{Slope} = \frac{40 \text{ km}}{5 \text{ h}}$$

Step 4 Divide the change in y by the change in x.

$$\text{Slope} = \frac{8 \text{ km}}{\text{h}}$$

The slope of the graph is 8 km/h.

Math Skill Handbook • **SR-27**

Bar Graph To compare data that does not change continuously you might choose a bar graph. A bar graph uses bars to show the relationships between variables. The *x*-axis variable is divided into parts. The parts can be numbers such as years, or a category such as a type of animal. The *y*-axis is a number and increases continuously along the axis.

Example

A recycling center collects 4.0 kg of aluminum on Monday, 1.0 kg on Wednesday, and 2.0 kg on Friday. Create a bar graph of this data.

Step 1 Select the *x*-axis and *y*-axis variables. The measured numbers (the masses of aluminum) should be placed on the *y*-axis. The variable divided into parts (collection days) is placed on the *x*-axis.

Step 2 Create a graph grid like you would for a line graph. Include labels and units.

Step 3 For each measured number, draw a vertical bar above the *x*-axis value up to the *y*-axis value. For the first data point, draw a vertical bar above Monday up to 4.0 kg.

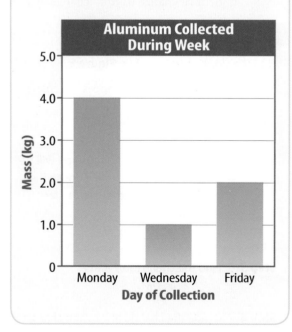

Practice Problem Draw a bar graph of the gases in air: 78% nitrogen, 21% oxygen, 1% other gases.

Circle Graph To display data as parts of a whole, you might use a circle graph. A circle graph is a circle divided into sections that represent the relative size of each piece of data. The entire circle represents 100%, half represents 50%, and so on.

Example

Air is made up of 78% nitrogen, 21% oxygen, and 1% other gases. Display the composition of air in a circle graph.

Step 1 Multiply each percent by 360° and divide by 100 to find the angle of each section in the circle.

$$78\% \times \frac{360°}{100} = 280.8°$$

$$21\% \times \frac{360°}{100} = 75.6°$$

$$1\% \times \frac{360°}{100} = 3.6°$$

Step 2 Use a compass to draw a circle and to mark the center of the circle. Draw a straight line from the center to the edge of the circle.

Step 3 Use a protractor and the angles you calculated to divide the circle into parts. Place the center of the protractor over the center of the circle and line the base of the protractor over the straight line.

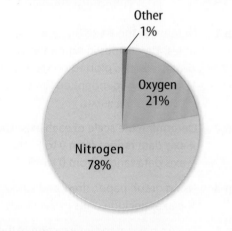

Practice Problem Draw a circle graph to represent the amount of aluminum collected during the week shown in the bar graph to the left.

Reference Handbook

Use and Care of a Microscope

Eyepiece Contains magnifying lenses you look through.

Arm Supports the body tube.

Low-power objective Contains the lens with the lowest power magnification.

Stage clips Hold the microscope slide in place.

Coarse adjustment Focuses the image under low power.

Fine adjustment Sharpens the image under high magnification.

Body tube Connects the eyepiece to the revolving nosepiece.

Revolving nosepiece Holds and turns the objectives into viewing position.

High-power objective Contains the lens with the highest magnification.

Stage Supports the microscope slide.

Light source Provides light that passes upward through the diaphragm, the specimen, and the lenses.

Base Provides support for the microscope.

Caring for a Microscope

1. Always carry the microscope holding the arm with one hand and supporting the base with the other hand.
2. Don't touch the lenses with your fingers.
3. The coarse adjustment knob is used only when looking through the lowest-power objective lens. The fine adjustment knob is used when the high-power objective is in place.
4. Cover the microscope when you store it.

Using a Microscope

1. Place the microscope on a flat surface that is clear of objects. The arm should be toward you.
2. Look through the eyepiece. Adjust the diaphragm so light comes through the opening in the stage.
3. Place a slide on the stage so the specimen is in the field of view. Hold it firmly in place by using the stage clips.
4. Always focus with the coarse adjustment and the low-power objective lens first. After the object is in focus on low power, turn the nosepiece until the high-power objective is in place. Use ONLY the fine adjustment to focus with the high-power objective lens.

Making a Wet-Mount Slide

1. Carefully place the item you want to look at in the center of a clean, glass slide. Make sure the sample is thin enough for light to pass through.
2. Use a dropper to place one or two drops of water on the sample.
3. Hold a clean coverslip by the edges and place it at one edge of the water. Slowly lower the coverslip onto the water until it lies flat.
4. If you have too much water or a lot of air bubbles, touch the edge of a paper towel to the edge of the coverslip to draw off extra water and draw out unwanted air.

Reference Handbook • SR-29

Diversity of Life: Classification of Living Organisms

A six-kingdom system of classification of organisms is used today. Two kingdoms—Kingdom Archaebacteria and Kingdom Eubacteria—contain organisms that do not have a nucleus and that lack membrane-bound structures in the cytoplasm of their cells. The members of the other four kingdoms have a cell or cells that contain a nucleus and structures in the cytoplasm, some of which are surrounded by membranes. These kingdoms are Kingdom Protista, Kingdom Fungi, Kingdom Plantae, and Kingdom Animalia.

Kingdom Archaebacteria

one-celled; some absorb food from their surroundings; some are photosynthetic; some are chemosynthetic; many are found in extremely harsh environments including salt ponds, hot springs, swamps, and deep-sea hydrothermal vents

Kingdom Eubacteria

one-celled; most absorb food from their surroundings; some are photosynthetic; some are chemosynthetic; many are parasites; many are round, spiral, or rod-shaped; some form colonies

Kingdom Protista

Phylum Euglenophyta one-celled; photosynthetic or take in food; most have one flagellum; euglenoids

Phylum Bacillariophyta one-celled; photosynthetic; have unique double shells made of silica; diatoms

Phylum Dinoflagellata one-celled; photosynthetic; contain red pigments; have two flagella; dinoflagellates

Phylum Chlorophyta one-celled, many-celled, or colonies; photosynthetic; contain chlorophyll; live on land, in freshwater, or salt water; green algae

Phylum Rhodophyta most are many-celled; photosynthetic; contain red pigments; most live in deep, saltwater environments; red algae

Phylum Phaeophyta most are many-celled; photosynthetic; contain brown pigments; most live in saltwater environments; brown algae

Phylum Rhizopoda one-celled; take in food; are free-living or parasitic; move by means of pseudopods; amoebas

Kingdom Eubacteria
Bacillus anthracis

Phylum Chlorophyta
Desmids

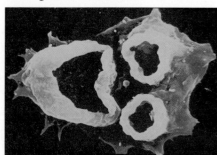

Amoeba

Phylum Zoomastigina one-celled; take in food; free-living or parasitic; have one or more flagella; zoomastigotes

Phylum Ciliophora one-celled; take in food; have large numbers of cilia; ciliates

Phylum Sporozoa one-celled; take in food; have no means of movement; are parasites in animals; sporozoans

Phyla Myxomycota and Acrasiomycota one- or many-celled; absorb food; change form during life cycle; cellular and plasmodial slime molds

Phylum Oomycota many-celled; are either parasites or decomposers; live in freshwater or salt water; water molds, rusts and downy mildews

Kingdom Fungi

Phylum Zygomycota many-celled; absorb food; spores are produced in sporangia; zygote fungi; bread mold

Phylum Ascomycota one- and many-celled; absorb food; spores produced in asci; sac fungi; yeast

Phylum Basidiomycota many-celled; absorb food; spores produced in basidia; club fungi; mushrooms

Phylum Deuteromycota members with unknown reproductive structures; imperfect fungi; *Penicillium*

Phylum Mycophycota organisms formed by symbiotic relationship between an ascomycote or a basidiomycote and green alga or cyanobacterium; lichens

Phylum Myxomycota
Slime mold

Phylum Oomycota
Phytophthora infestans

Lichens

Kingdom Plantae

Divisions Bryophyta (mosses), **Anthocerophyta** (hornworts), **Hepaticophyta** (liverworts), **Psilophyta** (whisk ferns) many-celled nonvascular plants; reproduce by spores produced in capsules; green; grow in moist, land environments

Division Lycophyta many-celled vascular plants; spores are produced in conelike structures; live on land; are photosynthetic; club mosses

Division Arthrophyta vascular plants; ribbed and jointed stems; scalelike leaves; spores produced in conelike structures; horsetails

Division Pterophyta vascular plants; leaves called fronds; spores produced in clusters of sporangia called sori; live on land or in water; ferns

Division Ginkgophyta deciduous trees; only one living species; have fan-shaped leaves with branching veins and fleshy cones with seeds; ginkgoes

Division Cycadophyta palmlike plants; have large, featherlike leaves; produces seeds in cones; cycads

Division Coniferophyta deciduous or evergreen; trees or shrubs; have needlelike or scalelike leaves; seeds produced in cones; conifers

Division Gnetophyta shrubs or woody vines; seeds are produced in cones; division contains only three genera; gnetum

Division Anthophyta dominant group of plants; flowering plants; have fruits with seeds

Kingdom Animalia

Phylum Porifera aquatic organisms that lack true tissues and organs; are asymmetrical and sessile; sponges

Phylum Cnidaria radially symmetrical organisms; have a digestive cavity with one opening; most have tentacles armed with stinging cells; live in aquatic environments singly or in colonies; includes jellyfish, corals, hydra, and sea anemones

Phylum Platyhelminthes bilaterally symmetrical worms; have flattened bodies; digestive system has one opening; parasitic and free-living species; flatworms

Division Bryophyta
Liverwort

Division Anthophyta
Tomato plant

Phylum Platyhelminthes
Flatworm

Phylum Chordata

Phylum Nematoda round, bilaterally symmetrical body; have digestive system with two openings; free-living forms and parasitic forms; roundworms

Phylum Mollusca soft-bodied animals, many with a hard shell and soft foot or footlike appendage; a mantle covers the soft body; aquatic and terrestrial species; includes clams, snails, squid, and octopuses

Phylum Annelida bilaterally symmetrical worms; have round, segmented bodies; terrestrial and aquatic species; includes earthworms, leeches, and marine polychaetes

Phylum Arthropoda largest animal group; have hard exoskeletons, segmented bodies, and pairs of jointed appendages; land and aquatic species; includes insects, crustaceans, and spiders

Phylum Echinodermata marine organisms; have spiny or leathery skin and a water-vascular system with tube feet; are radially symmetrical; includes sea stars, sand dollars, and sea urchins

Phylum Chordata organisms with internal skeletons and specialized body systems; most have paired appendages; all at some time have a notochord, nerve cord, gill slits, and a post-anal tail; include fish, amphibians, reptiles, birds, and mammals

PERIODIC TABLE OF THE ELEMENTS

Key:
- Element — Hydrogen
- Atomic number — 1
- Symbol — H
- Atomic mass — 1.01
- State of matter

Legend:
- 🎈 Gas
- 💧 Liquid
- ▫ Solid
- ⊙ Synthetic

A column in the periodic table is called a **group**.

A row in the periodic table is called a **period**.

Group	1	2	3	4	5	6	7	8	9
1	Hydrogen 1 H 1.01								
2	Lithium 3 Li 6.94	Beryllium 4 Be 9.01							
3	Sodium 11 Na 22.99	Magnesium 12 Mg 24.31							
4	Potassium 19 K 39.10	Calcium 20 Ca 40.08	Scandium 21 Sc 44.96	Titanium 22 Ti 47.87	Vanadium 23 V 50.94	Chromium 24 Cr 52.00	Manganese 25 Mn 54.94	Iron 26 Fe 55.85	Cobalt 27 Co 58.93
5	Rubidium 37 Rb 85.47	Strontium 38 Sr 87.62	Yttrium 39 Y 88.91	Zirconium 40 Zr 91.22	Niobium 41 Nb 92.91	Molybdenum 42 Mo 95.96	Technetium 43 Tc (98)	Ruthenium 44 Ru 101.07	Rhodium 45 Rh 102.91
6	Cesium 55 Cs 132.91	Barium 56 Ba 137.33	Lanthanum 57 La 138.91	Hafnium 72 Hf 178.49	Tantalum 73 Ta 180.95	Tungsten 74 W 183.84	Rhenium 75 Re 186.21	Osmium 76 Os 190.23	Iridium 77 Ir 192.22
7	Francium 87 Fr (223)	Radium 88 Ra (226)	Actinium 89 Ac (227)	Rutherfordium 104 Rf (267)	Dubnium 105 Db (268)	Seaborgium 106 Sg (271)	Bohrium 107 Bh (272)	Hassium 108 Hs (270)	Meitnerium 109 Mt (276)

The number in parentheses is the mass number of the longest lived isotope for that element.

Lanthanide series:

Cerium 58 Ce 140.12	Praseodymium 59 Pr 140.91	Neodymium 60 Nd 144.24	Promethium 61 Pm (145)	Samarium 62 Sm 150.36	Europium 63 Eu 151.96

Actinide series:

Thorium 90 Th 232.04	Protactinium 91 Pa 231.04	Uranium 92 U 238.03	Neptunium 93 Np (237)	Plutonium 94 Pu (244)	Americium 95 Am (243)

Metal
Metalloid
Nonmetal
Recently discovered

			13	14	15	16	17	18
								Helium 2 He 4.00
			Boron 5 B 10.81	Carbon 6 C 12.01	Nitrogen 7 N 14.01	Oxygen 8 O 16.00	Fluorine 9 F 19.00	Neon 10 Ne 20.18
10	11	12	Aluminum 13 Al 26.98	Silicon 14 Si 28.09	Phosphorus 15 P 30.97	Sulfur 16 S 32.07	Chlorine 17 Cl 35.45	Argon 18 Ar 39.95
Nickel 28 Ni 58.69	Copper 29 Cu 63.55	Zinc 30 Zn 65.38	Gallium 31 Ga 69.72	Germanium 32 Ge 72.64	Arsenic 33 As 74.92	Selenium 34 Se 78.96	Bromine 35 Br 79.90	Krypton 36 Kr 83.80
Palladium 46 Pd 106.42	Silver 47 Ag 107.87	Cadmium 48 Cd 112.41	Indium 49 In 114.82	Tin 50 Sn 118.71	Antimony 51 Sb 121.76	Tellurium 52 Te 127.60	Iodine 53 I 126.90	Xenon 54 Xe 131.29
Platinum 78 Pt 195.08	Gold 79 Au 196.97	Mercury 80 Hg 200.59	Thallium 81 Tl 204.38	Lead 82 Pb 207.20	Bismuth 83 Bi 208.98	Polonium 84 Po (209)	Astatine 85 At (210)	Radon 86 Rn (222)
Darmstadtium 110 Ds (281)	Roentgenium 111 Rg (280)	Copernicium 112 Cn (285)	Ununtrium * 113 Uut (284)	Ununquadium * 114 Uuq (289)	Ununpentium * 115 Uup (288)	Ununhexium * 116 Uuh (293)		Ununoctium * 118 Uuo (294)

* The names and symbols for elements 113-116 and 118 are temporary. Final names will be selected when the elements' discoveries are verified.

Gadolinium 64 Gd 157.25	Terbium 65 Tb 158.93	Dysprosium 66 Dy 162.50	Holmium 67 Ho 164.93	Erbium 68 Er 167.26	Thulium 69 Tm 168.93	Ytterbium 70 Yb 173.05	Lutetium 71 Lu 174.97
Curium 96 Cm (247)	Berkelium 97 Bk (247)	Californium 98 Cf (251)	Einsteinium 99 Es (252)	Fermium 100 Fm (257)	Mendelevium 101 Md (258)	Nobelium 102 No (259)	Lawrencium 103 Lr (262)

Glossary/Glosario

Multilingual eGlossary

A science multilingual glossary is available on the science Web site. The glossary includes the following languages.

Arabic	Hmong	Tagalog
Bengali	Korean	Urdu
Chinese	Portuguese	Vietnamese
English	Russian	
Haitian Creole	Spanish	

Cómo usar el glosario en español:
1. Busca el término en inglés que desees encontrar.
2. El término en español, junto con la definición, se encuentran en la columna de la derecha.

Pronunciation Key

Use the following key to help you sound out words in the glossary.

a	back (BAK)		**ew**	food (FEWD)
ay	day (DAY)		**yoo**	pure (PYOOR)
ah	father (FAH thur)		**yew**	few (FYEW)
ow	flower (FLOW ur)		**uh**	comma (CAH muh)
ar	car (CAR)		**u (+ con)**	rub (RUB)
e	less (LES)		**sh**	shelf (SHELF)
ee	leaf (LEEF)		**ch**	nature (NAY chur)
ih	trip (TRIHP)		**g**	gift (GIHFT)
i (i + con + e)	idea (i DEE uh)		**j**	gem (JEM)
oh	go (GOH)		**ing**	sing (SING)
aw	soft (SAWFT)		**zh**	vision (VIH zhun)
or	orbit (OR buht)		**k**	cake (KAYK)
oy	coin (COYN)		**s**	seed, cent (SEED, SENT)
oo	foot (FOOT)		**z**	zone, raise (ZOHN, RAYZ)

English / Español — A

active transport/asexual reproduction — **transporte activo/reproducción asexual**

active transport: the movement of substances through a cell membrane using the cell's energy. (p. 64)

adaptation (a dap TAY shun): an inherited trait that increases an organism's chance of surviving and reproducing in a particular environment. (p. 203)

allele (uh LEEL): a different form of a gene. (p. 160)

analogous (uh NAH luh gus) structures: body parts that perform a similar function but differ in structure. (p. 211)

asexual reproduction: a type of reproduction in which one parent organism produces offspring without meiosis and fertilization. (p. 129)

transporte activo: movimiento de sustancias a través de la membrana celular usando la energía de la célula. (pág. 64)

adaptación: rasgo heredado que aumenta la oportunidad de un organismo de sobrevivir y reproducirse en su medioambiente. (pág. 203)

alelo: forma diferente de un gen. (pág. 160)

estructuras análogas: partes del cuerpo que ejecutan una función similar pero tienen una estructura distinta. (pág. 211)

reproducción asexual: tipo de reproducción en la cual un organismo parental produce crías sin mitosis ni fertilización. (pág. 129)

B

binomial nomenclature: a naming system that gives each organism a two-word scientific name. (p. 21)

biological evolution: the change over time in populations of related organisms. (p. 195)

budding: the process during which a new organism grows by mitosis and cell division on the body of its parent. (p. 131)

nomenclatura binomial: sistema de nombrar que le da a cada organismo un nombre científico de dos palabras. (pág. 21)

evolución biológica: cambio a través del tiempo en las poblaciones de organismos relacionados. (pág. 195)

germinación: proceso durante el cual un organismo nuevo crece por medio de mitosis y división celular en el cuerpo de su progenitor. (pág. 131)

C

camouflage (KAM uh flahj): an adaptation that enables a species to blend in with its environment. (p. 204)

carbohydrate (kar boh HI drayt): a macromolecule made up of one or more sugar molecules, which are composed of carbon, hydrogen, and oxygen; usually the body's major source of energy. (p. 47)

cast: a fossil copy of an organism made when a mold of the organism is filled with sediment or mineral deposits. (p. 191)

cell: the smallest unit of life. (p. 10)

cell cycle: a cycle of growth, development, and division that most cells in an organism go through. (p. 85)

cell differentiation (dihf uh ren shee AY shun): the process by which cells become different types of cells. (p. 99)

cell membrane: a flexible covering that protects the inside of a cell from the environment outside the cell. (p. 52)

cell theory: the theory that states that all living things are made of one or more cells, the cell is the smallest unit of life, and all new cells come from preexisting cells. (p. 44)

cell wall: a stiff structure outside the cell membrane that protects a cell from attack by viruses and other harmful organisms. (p. 52)

camuflaje: adaptación que permite a las especies mezclarse con su medioambiente. (pág. 204)

carbohidrato: macromolécula constituida de una o más moléculas de azúcar, las cuales están compuestas de carbono, hidrógeno y oxígeno; usualmente es la mayor fuente de energía del cuerpo. (pág. 47)

contramolde: copia fósil de un organismo compuesto en un molde de el organismo está lleno de sedimentos o los depósitos de minerales. (pág. 191)

célula: unidad más pequeña de vida. (pág. 10)

ciclo celular: ciclo de crecimiento, desarrollo y división por el que pasan la mayoría de células de un organismo. (pág. 85)

diferenciación celular: proceso por el cual las células se convierten en diferentes tipos de células. (pág. 99)

membrana celular: cubierta flexible que protege el interior de una célula del ambiente externo de la célula. (pág. 52)

teoría celular: teoría que establece que todos los seres vivos están constituidos de una o más células (la célula es la unidad más pequeña de vida) y que las células nuevas provienen de células preexistentes. (pág. 44)

pared celular: estructura rígida en el exterior de la membrana celular que protege la célula del ataque de virus y otros organismos dañinos. (pág. 52)

cellular respiration/dichotomous key

cellular respiration: a series of chemical reactions that convert the energy in food molecules into a usable form of energy called ATP. (p. 69)

centromere: a structure that holds sister chromatids together. (p. 88)

chloroplast (KLOR uh plast): a membrane-bound organelle that uses light energy and makes food—a sugar called glucose—from water and carbon dioxide in a process known as photosynthesis. (p. 57)

cladogram: a branched diagram that shows the relationships among organisms, including common ancestors. (p. 23)

cloning: a type of asexual reproduction performed in a laboratory that produces identical individuals from a cell or a cluster of cells taken from a multicellular organism. (p. 134)

codominance: an inheritance pattern in which both alleles can be observed in a phenotype. (p. 164)

comparative anatomy: the study of similarities and differences among structures of living species. (p. 210)

compound microscope: a light microscope that uses more than one lens to magnify an object. (p. 28)

cytokinesis (si toh kuh NEE sus): a process during which the cytoplasm and its contents divide. (p. 89)

cytoplasm: the liquid part of a cell inside the cell membrane; contains salts and other molecules. (p. 53)

cytoskeleton: a network of threadlike proteins joined together that gives a cell its shape and helps it move. (p. 53)

respiración celular/clave dicotómica

respiración celular: serie de reacciones químicas que convierten la energía de las moléculas de alimento en una forma de energía utilizable llamada ATP. (pág. 69)

centrómero: estructura que mantiene unidas las cromátidas hermanas. (pág. 88)

cloroplasto: organelo limitado por una membrana que usa la energía lumínica para producir alimento –un azúcar llamado glucosa– del agua y del dióxido de carbono en un proceso llamado fotosíntesis. (pág. 57)

cladograma: diagrama de brazos que muestra las relaciones entre los organismos, incluidos los ancestros comunes. (pág. 23)

clonación: tipo de reproducción asexual realizada en un laboratorio que produce individuos idénticos a partir de una célula o grupo de células tomadas de un organismo pluricelular. (pág. 134)

condominante: patrón heredado en el cual los dos alelos se observan en un fenotipo. (pág. 164)

anatomía comparativa: estudio de las similitudes y diferencias entre las estructuras de las especies vivas. (pág. 210)

microscopio compuesto: microscopio de luz que usa más de un lente para aumentar la imagen de un objeto. (pág. 28)

citocinesis: proceso durante el cual el citoplasma y sus contenidos se dividen. (pág. 89)

citoplasma: fluido en el interior de una célula que contiene sales y otras moléculas. (pág. 53)

citoesqueleto: red de proteínas en forma de filamentos unidos que le da forma a la célula y le ayuda a moverse. (pág. 53)

D

daughter cells: the two new cells that result from mitosis and cytokinesis. (p. 89)

dichotomous key: a series of descriptions arranged in pairs that leads the user to the identification of an unknown organism. (p. 22)

células hija: las dos células nuevas que resultan de la mitosis y la citocinesis. (pág. 89)

clave dicotómica: serie de descripciones organizadas en pares que dan al usuario la identificación de un organismo desconocido. (pág. 22)

diffusion/fossil record

diffusion: the movement of substances from an area of higher concentration to an area of lower concentration. (p. 62)

diploid: a cell that has pairs of chromosomes. (p. 118)

DNA: the abbreviation for deoxyribonucleic (dee AHK sih ri boh noo klee ihk) acid, an organism's genetic material. (p. 170)

dominant (DAH muh nunt) trait: a genetic factor that blocks another genetic factor. (p. 155)

E

egg: the female reproductive, or sex, cell; forms in an ovary. (p. 117)

electron microscope: a microscope that uses a magnetic field to focus a beam of electrons through an object or onto an object's surface. (p. 29)

embryology (em bree AH luh jee): the science of the development of embryos from fertilization to birth. (p. 212)

endocytosis (en duh si TOH sus): the process during which a cell takes in a substance by surrounding it with the cell membrane. (p. 64)

exocytosis (ek soh si TOH sus): the process during which a cell's vesicles release their contents outside the cell. (p. 64)

extinction (ihk STINGK shun): event that occurs when the last individual organism of a species dies. (p. 194)

F

facilitated diffusion: the process by which molecules pass through a cell membrane using special proteins called transport proteins. (p. 63)

fermentation: a reaction that eukaryotic and prokaryotic cells can use to obtain energy from food when oxygen levels are low. (p. 70)

fertilization (fur tuh luh ZAY shun): a reproductive process in which a sperm joins with an egg. (p. 117)

fission: cell division that forms two genetically identical cells. (p. 130)

fossil record: record of all the fossils ever discovered on Earth. (p. 189)

difusión/registro fósil

difusión: movimiento de sustancias de un área de mayor concentración a un área de menor concentración. (pág. 62)

diploide: célula que tiene pares de cromosomas. (pág. 118)

ADN: abreviatura para ácido desoxirribonucleico, material genético de un organismo. (pág. 170)

rasgo dominante: factor genético que bloquea otro factor genético. (pág. 155)

óvulo: célula reproductiva femenina o sexual; forma en un ovario. (pág. 117)

microscopio electrónico: microscopio que usa un campo magnético para enfocar un haz de electrones a través de un objeto o sobre la superficie de un objeto. (pág. 29)

embriología: ciencia que trata el desarrollo de embriones desde la fertilización hasta el nacimiento. (pág. 212)

endocitosis: proceso durante el cual una célula absorbe una sustancia rodeándola con la membrana celular. (pág. 64)

exocitosis: proceso durante el cual las vesículas de una célula liberan sus contenidos fuera de la célula. (pág. 64)

extinción: evento que ocurre cuando el último organismo individual de una especie muere. (pág. 194)

difusión facilitada: proceso por el cual las moléculas pasan a través de la membrana celular usando proteínas especiales, llamadas proteínas de transporte. (pág. 63)

fermentación: reacción que las células eucarióticas y procarióticas usan para obtener energía del alimento cuando los niveles de oxígeno son bajos. (pág. 70)

fertilización: proceso reproductivo en el cual un espermatozoide se une con un óvulo. (pág. 117)

fisión: división celular que forma dos células genéticamente idénticas. (pág. 130)

registro fósil: registro de todos los fósiles descubiertos en la Tierra. (pág. 189)

G

gene (JEEN): a section of DNA on a chromosome that has genetic information for one trait. (p. 160)

genetics: the study of how traits are passed from parents to offspring. (p. 149)

genotype (JEE nuh tipe): the alleles of all the genes on an organism's chromosomes; controls an organism's phenotype. (p. 160)

genus (JEE nus): a group of similar species. (p. 21)

geologic time scale: a chart that divides Earth's history into different time units based on changes in the rocks and fossils. (p. 193)

glycolysis: a process by which glucose, a sugar, is broken down into smaller molecules. (p. 69)

gen: parte del ADN en un cromosoma que contiene información genética para un rasgo. (pág. 160)

genética: estudio de cómo los rasgos pasan de los padres a los hijos. (pág. 149)

genotipo: de los alelos de todos los genes en los cromosomas de un organismo, los controles de fenotipo de un organismo. (pág. 160)

género: grupo de especies similares. (pág. 21)

escala de tiempo geológico: tabla que divide la historia de la Tierra en diferentes unidades de tiempo, basado en los cambios en las rocas y fósiles. (pág. 193)

glucólisis: proceso por el cual la glucosa, un azúcar, se divide en moléculas más pequeñas. (pág. 69)

H

haploid: a cell that has only one chromosome from each pair. (p. 119)

heredity (huh REH duh tee): the passing of traits from parents to offspring. (p. 149)

heterozygous (he tuh roh ZI gus): a genotype in which the two alleles of a gene are different. (p. 161)

homeostasis (hoh mee oh STAY sus): an organism's ability to maintain steady internal conditions when outside conditions change. (p. 13)

homologous (huh MAH luh gus) chromosomes: pairs of chromosomes that have genes for the same traits arranged in the same order. (p. 118)

homologous (huh MAH luh gus) structures: body parts of organisms that are similar in structure and position but different in function. (p. 210)

homozygous (hoh muh ZI gus): a genotype in which the two alleles of a gene are the same. (p. 161)

haploide: célula que tiene solamente un cromosoma de cada par. (pág. 119)

herencia: paso de rasgos de los padres a los hijos. (pág. 149)

heterocigoto: genotipo en el cual los dos alelos de un gen son diferentes. (pág. 161)

homeostasis: capacidad de un organismo de mantener las condiciones internas estables cuando las condiciones externas cambian. (pág. 13)

cromosomas homólogos: pares de cromosomas que tienen genes de iguales rasgos dispuestos en el mismo orden. (pág. 118)

estructuras homólogas: partes del cuerpo de los organismos que son similares en estructura y posición pero diferentes en función. (pág. 210)

homocigoto: genotipo en el cual los dos alelos de un gen son iguales. (pág. 161)

I

incomplete dominance: an inheritance pattern in which an offspring's phenotype is a combination of the parents' phenotypes. (p. 164)

dominancia incompleta: patrón heredado en el cual el fenotipo de un hijo es una combinación de los fenotipos de los padres. (pág. 164)

interphase/nucleotide

interphase: the period during the cell cycle of a cell's growth and development. (p. 86)

interfase/nucelótido

interfase: período durante el ciclo celular del crecimiento y desarrollo de una célula. (pág. 86)

light microscope: a microscope that uses light and lenses to enlarge an image of an object. (p. 28)

lipid: a large macromolecule that does not dissolve in water. (p. 47)

microscopio de luz: microscopio que usa luz y lentes para aumentar la imagen de un objeto. (pág. 28)

lípido: macromolécula extensa que no se disuelve en agua. (pág. 47)

macromolecule: substance that forms from joining many small molecules together. (p. 45)

meiosis: a process in which one diploid cell divides to make four haploid sex cells. (p. 119)

mimicry (MIH mih kree): an adaptation in which one species looks like another species. (p. 204)

mitosis (mi TOH sus): a process during which the nucleus and its contents divide. (p. 89)

mold: the impression of an organism in a rock. (p. 191)

multicellular: a living thing that is made up of two or more cells. (p. 10)

mutation (myew TAY shun): a permanent change in the sequence of DNA, or the nucleotides, in a gene or a chromosome. (p. 175)

macromolécula: sustancia que se forma al unir muchas moléculas pequeñas. (pág. 45)

meiosis: proceso en el cual una célula diploide se divide para constituir cuatro células sexuales haploides. (pág. 119)

mimetismo: una adaptación en el cual una especie se parece a otra especie. (pág. 204)

mitosis: proceso durante el cual el núcleo y sus contenidos se divide. (pág. 89)

molde: impresión de un organismo en una roca. (pág. 191)

pluricelular: ser vivo formado por dos o más células. (pág. 10)

mutación: cambio permanente en la secuencia de ADN, de los nucleótidos, en un gen o en un cromosoma. (pág. 175)

naturalist: a person who studies plants and animals by observing them. (p. 199)

natural selection: the process by which organisms with variations that help them survive in their environment live longer, compete better, and reproduce more than those that do not have the variations. (p. 202)

nucleic acid: a macromolecule that forms when long chains of molecules called nucleotides join together. (p. 46)

nucleotide (NEW klee uh tide): a molecule made of a nitrogen base, a sugar, and a phosphate group. (p. 171)

naturalista: persona que estudia las plantas y los animales por medio de la observación. (pág. 199)

selección natural: proceso por el cual los organismos con variaciones que las ayudan a sobrevivir en sus medioambientes viven más, compiten mejor y se reproducen más que aquellas que no tienen esas variaciones. (pág. 202)

ácido nucléico: macromolécula que se forma cuando cadenas largas de moléculas llamadas nucleótidos se unen. (pág. 46)

nucelótido: molécula constituida de una base de nitrógeno, azúcar y un grupo de fosfato. (pág. 171)

nucleus/recessive trait

nucleus: part of a eukaryotic cell that directs cell activity and contains genetic information stored in DNA. (p. 55)

núcleo/rasgo recesivo

núcleo: parte de la célula eucariótica que gobierna la actividad celular y contiene la información genética almacenada en el ADN. (pág. 55)

O

organ: a group of different tissues working together to perform a particular job. (p. 102)

organelle: membrane-surrounded component of a eukaryotic cell with a specialized function. (p. 54)

organism: something that has all the characteristics of life. (p. 9)

organ system: a group of organs that work together and perform a specific task. (p. 103)

osmosis: the diffusion of water molecules only through a membrane. (p. 62)

órgano: grupo de diferentes tejidos que trabajan juntos para realizar una función específica. (pág. 102)

organelo: componente de una célula eucariótica rodeado de una membrana con una función especializada. (pág. 54)

organismo: algo que tiene todas las características de la vida. (pág. 9)

sistema de órganos: grupo de órganos que trabajan juntos y realizar una función específica. (pág. 103)

ósmosis: difusión de las moléculas de agua únicamente a través de una membrana. (pág. 62)

P

passive transport: the movement of substances through a cell membrane without using the cell's energy. (p. 61)

phenotype (FEE nuh tipe): how a trait appears or is expressed. (p. 160)

photosynthesis: a series of chemical reactions that convert light energy, water, and CO_2 into the food-energy molecule glucose and give off oxygen. (p. 71)

polygenic inheritance: an inheritance pattern in which multiple genes determine the phenotype of a trait. (p. 165)

protein: a long chain of amino acid molecules; contains carbon, hydrogen, oxygen, nitrogen, and sometimes sulfur. (p. 47)

Punnett square: a model that is used to show the probability of all possible genotypes and phenotypes of offspring. (p. 162)

transporte pasivo: movimiento de sustancias a través de una membrana celular sin usar la energía de la célula. (pág. 61)

fenotipo: forma como aparece o se expresa un rasgo. (pág. 160)

fotosíntesis: serie de reacciones químicas que convierten la energía lumínica, el agua y el CO_2 en glucosa, una molécula de energía alimentaria, y libera oxígeno. (pág. 71)

herencia poligénica: patrón de herencia en el cual genes múltiples determinan el fenotipo de un rasgo. (pág. 165)

proteína: larga cadena de aminoácidos; contiene carbono, hidrógeno, oxígeno, nitrógeno y, algunas veces, sulfuro. (pág. 47)

cuadro de Punnett: modelo que se utiliza para demostrar la probabilidad de que todos los genotipos y fenotipos posibles de cría. (pág. 162)

R

recessive (rih SE sihv) trait: a genetic factor that is blocked by the presence of a dominant factor. (p. 155)

rasgo recesivo: factor genético boqueado por la presencia de un factor dominante. (pág. 155)

regeneration/unicellular

regeneration: a type of asexual reproduction that occurs when an offspring grows from a piece of its parent. (p. 132)

replication: the process of copying a DNA molecule to make another DNA molecule. (p. 172)

RNA: ribonucleic acid, a type of nucleic acid that carries the code for making proteins from the nucleus to the cytoplasm. (p. 173)

regeneración/unicelular

regeneración: tipo de reproducción asexual que ocurre cuando un organismo se origina de una parte de su progenitor. (pág. 132)

replicación: proceso por el cual se copia una molécula de ADN para hacer otra molécula de ADN. (pág. 172)

ARN: ácido ribonucleico, un tipo de ácido nucléico que contiene el código para hacer proteínas del núcleo para el citoplasma. (pág. 173)

S

selective breeding: the selection and breeding of organisms for desired traits. (p. 205)

sexual reproduction: a type of reproduction in which the genetic material from two different cells—a sperm and an egg—combine, producing an offspring. (p. 117)

sister chromatids: two identical chromosomes that make up a duplicated chromosome. (p. 88)

species (SPEE sheez): a group of organisms that have similar traits and are able to produce fertile offspring. (p. 21)

sperm: a male reproductive, or sex, cell; forms in a testis. (p. 117)

stem cell: an unspecialized cell that is able to develop into many different cell types. (p. 100)

cría selectiva: selección y la cría de organismos para las características deseadas. (pág. 205)

reproducción sexual: tipo de reproducción en la cual el material genético de dos células diferentes de un espermatozoide y un óvulo se combinan, produciendo una cría. (pág. 117)

cromátidas hermanas: dos cromosomas idénticos que constituyen un cromosoma duplicado. (pág. 88)

especie: grupo de organismos que tienen rasgos similares y que están en capacidad de producir crías fértiles. (pág. 21)

esperma: célula reproductora masculina o sexual; forma en un testículo. (pág. 117)

célula madre: célula no especializada que tiene la capacidad de desarrollarse en diferentes tipos de células. (pág. 100)

T

tissue: a group of similar types of cells that work together to carry out specific tasks. (p. 101)

trace fossil: the preserved evidence of the activity of an organism. (p. 191)

transcription: the process of making mRNA from DNA. (p. 173)

translation: the process of making a protein from RNA. (p. 174)

tejido: grupo de tipos similares de células que trabajan juntas para llevar a cabo diferentes funciones. (pág. 101)

traza fósil: evidencia conservada de la actividad de un organismo. (pág. 191)

transcripción: proceso por el cual se hace mARN de ADN. (pág. 173)

traslación: proceso por el cual se hacen proteínas a partir de ARN. (pág. 174)

U

unicellular: a living thing that is made up of only one cell. (p. 10)

unicelular: ser vivo formado por una sola célula. (pág. 10)

variation/zygote **variación/zigoto**

V

variation (ver ee AY shun): a slight difference in an inherited trait among individual members of a species. (p. 201)

vegetative reproduction: a form of asexual reproduction in which offspring grow from a part of a parent plant. (p. 133)

vestigial (veh STIH jee ul) structure: body part that has lost its original function through evolution. (p. 211)

variación: ligera diferencia en un rasgo hereditario entre los miembros individuales de una especie. (pág. 201)

reproducción vegetativa: forma de reproducción asexual en la cual el organismo se origina a partir de una planta parental. (pág. 133)

estructura vestigial: parte del cuerpo que a través de la evolución perdió la función original. (pág. 211)

Z

zygote (ZI goht): the new cell that forms when a sperm cell fertilizes an egg cell. (p. 117)

zigoto: célula nueva que se forma cuando un espermatozoide fecunda un óvulo. (pág. 117)

Index

Italic numbers = illustration/photo **Bold numbers** = vocabulary term
lab = indicates entry is used in a lab on this page

A

ABO blood types
  explanation of, 165, *165*
Academic Vocabulary, 30, **52**, 102, **132**, 165. See also **Vocabulary**
Active transport
  explanation of, *64*, **64**
Adenine
  in nucleotide, 171
Adenosine triphosphate (ATP)
  explanation of, 56, 69
  glucose converted into, 71
  use of, 70
Air-breathing catfish, 17
Alcohol fermentation
  explanation of, 71
Allele(s)
  dominant and recessive, 161, *161*, *162*, *162*, 163, *163*
  explanation of, *160*, **160**
  multiple, 165
Amino acid(s)
  proteins made from, 174
Amoeba(s)
  explanation of, 98, *98*
  reproduction in, 131, *131*
Amylase, 47
Anaphase, 120, *120*, 121, *121*
  explanation of, 91, *91*
Animal cell(s)
  plant cells v., 52, 59
Animal cloning
  explanation of, *135*, **135**
Animal regeneration
  explanation of, 132, *132*
Animalia, 20
Appendage(s)
  cell, 53, *53*
Aquino, Adriana, 17
Arachnologist(s), 127
Araschnia levana, 166
Archaea, 20, *20*
Aristotle, 19
Armored catfish, 17
Asexual reproduction
  advantages of, 136
  animal regeneration as, 132, *132*
  budding as, 131, *131*
  cloning as, *134*, 134–135, *135*
  disadvantages of, 136
  explanation of, **129**, 140
  fission as, 130, *130*
  mitotic cell division as, 131, *131*
  vegetative, 133, *133*, 133 *lab*
Atomic force microscope (AFM), 49
Atoms
  explanation of, **29**

B

Bacteria
  classification of, 21, *21*
  reproduction in, 130
Bateson, William, 157
Big Idea, 6, 34, 40, 76, 82, 108, 114, 140, 146, 180
  Review, 37, 79, 111, 143, 183
Binomial nomenclature
  explanation of, **21**
Blending inheritance
  explanation of, 149
Body temperature
  regulation of, 13
Breast cancer
  explanation of, *176*
Brown bear
  classification of, *21*, 22
Budding
  explanation of, *131*, 131

C

Carbohydrate(s)
  explanation of, 46, *46*, **47**
Carbon dioxide
  alcohol fermentation producing, 71, *71*
  in photosynthesis, 70, 72, *72*
Careers in Science, 17, 127
Carnivora, 21
Carrier proteins, 63
Catfish
  air-breathing, 17
Cell cycle
  cytokinesis and, 92, *92*
  explanation of, 85, *85*
  interphase of, *86*, 86–88, *87*, 88
  length of, 87, *87*
  mitotic phase of, 86, *88*, 89, *90*, 90–92, *91*, *92*
  organelle replication and, 89, *89*
  phases of, 86, *86*, 88, *88*
  results of, 92–93
Cell differentiation
  in eukaryotes, 99
  explanation of, **99**
  function of, 106–107 *lab*
Cell division
  comparison of types of, *123*
Cell division
  in animal cells, 92
  explanation of, 99
  organelles during, 89
  in humans, 92, 93
  in plant cells, 92
  results of, 92–93

Cell membrane
  active transport in, 64, *64*
  diffusion in, 62, *63*
  explanation of, **52**
  function of, 61, 61 *lab*
  substances in, 63 *lab*
Cell organelle(s)
  explanation of, 55
  manufacturing molecules in, 56
  nucleus as, 55, *55*
  processing, transporting and storing molecules in, 57, *57*
  processing energy in, 56–57
Cell theory
  explanation of, *44*, **44**
Cell wall
  explanation of, 52, *52*
Cell(s)
  animal stem, 100
  appendages to, 53, *53*
  communication between, 47
  composition of, 43 *lab*
  cytoplasm in, 53
  cytoskeleton in, 53
  discovery of, 43, *43*
  energy processing in, 56–57
  energy use by, 14
  eukaryotic, 54, *54*
  explanation of, **10**, 97
  features of, 44, *44*
  functions of, 52, 65, 76
  macromolecules in, 46, 46–47, 47 *lab*
  molecules in, 170
  nucleus of, 55, *55*
  passive transport in, 61
  plant, 100, *100*, 104
  production of, 85 *lab*
  prokaryotic, 54, *54*
  proteins in, 47, *47*, 56, 57
  role of, 103 *lab*
  size and shape of, 51, *51*, 65, 67
  water in, 45, *45*
Cellular respiration
  explanation of, 69
  glycolysis in, 69, *69*
  reactions in mitochondria in, 70, *70*
  relationship between photosynthesis and, 72, *72*
Centromere
  explanation of, 88, *88*
  meiosis and, 120
Channel proteins
  explanation of, 63
Chapter Review, 36–37, 78–79, 110–111, 142–143, 182–183
Chargaff, Erwin, 171
Chase, Martha, 157

I-2 • Index

Chloroplast(s)
  explanation of, *57,* **57,** 89
  reactions in, 72
Cholesterol
  explanation of, 47
Chromatid(s)
  as chromosomes, 89
  sister, 88, *88,* 89
Chromatin
  explanation of, 87
  replicated, 90, *90*
Chromosome(s)
  cell division and, 88, *88*
Chromosome(s)
  composition of, 170
  DNA in, *170,* 170–172
  explanation of, 55, *118,* **118**
  genetic information in, 159, 160, *160*
  homologous, 118
  human, 159, *159,* 173, 175
  Mendel's laws of inheritance and, 157
  in sex cells, 119, 122
Cilia
  explanation of, 53, *53*
Cladogram(s)
  explanation of, 23, *23*
Cloning
  animal, 135, *135*
  explanation of, **134**
  plant, 134, *134*
CODIS (Combined DNA Index System), 95
Codominance
  explanation of, *164,* **164**
Codon(s)
  explanation of, 174
Common Use. *See* **Science Use v. Common Use**
Complex
  explanation of, **102**
Compound microscope
  explanation of, *28,* **28**
Conclude
  explanation of, **165**
Connective tissue
  in animals, 101, 102
Contractile vacuole, 98, *98*
  explanation of, 13, *13*
Crab(s), 132
Crabgrass
  reproduction in, 136, *136*
Crick, Francis, 157
Critical thinking, 16, 24, 31, 37, 48, 58, 66, 73, 79, 94, 105, 116, 137, 143, 156, 167, 177
Cross-pollination
  experiments in, *151,* 151–153, *152, 153*
  explanation of, 150
Culture
  explanation of, **134**
Cystic fibrosis
  explanation of, *176*
Cytokinesis
  explanation of, **89,** 92, *92,* 119, 131
Cytoplasm
  during cytokinesis, 89, 92
  explanation of, **53,** 86

in meiosis, 119, 120, *120*
  in mitosis, 119, 131
Cytosine
  in nucleotide, 171
Cytoskeleton
  explanation of, *53,* **53**

D

Daughter cells
  explanation of, 82, **89**
Deletion mutation
  explanation of, 175, *175*
Deoxyribonucleic acid (DNA)
  explanation of, 46, 55
  observation of, 47 *lab*
Dermal tissue
  in plants, 101, 102, *102*
Dichotomous key(s)
  construction of, 32–33 *lab*
  explanation of, *22,* **22**
  use of, 25
Diffusion
  explanation of, *62,* **62**
  facilitated, 63, *63*
Digestive system
  human, 103
Diploid cell(s)
  chromosomes of, 118, 119
  explanation of, **118**
  meiosis in, 119, 119 *lab,* 122, *122*
DNA
  cell cycle and, 87, *87,* 88, *88,* 95
  explanation of, **124,** *170,* **170**
  in human cells, 95
  junk, 173
  making mRNA from, 173, *173*
  modeling of, 172 *lab*
  mutations and, *175,* 175–176
  in prokaryotes, 130
  replication of, 172, *172*
  research on, 157
  sexual reproduction and, 124, 129
  structure of, 171, *171,* 173
Dolly (sheep), 135, *135*
Domain(s)
  explanation of, 20, *20*
Dominance
  types of, 164, *164*
Dominant allele(s)
  explanation of, *161,* 161–163
Dominant trait(s)
  explanation of, **155**
  identification of, 155 *lab*
Down syndrome
  explanation of, 118

E

E. coli
  reproduction in, 130
Egg cells
  explanation of, **117, 150**
Electron microscope
  explanation of, *29,* **29,** 30
Endangered species
  cloning of, 135

Endocytosis
  explanation of, *64,* **64**
Endoplasmic reticulum
  explanation of, 56, *56*
Energy use
  by organisms, 14, *14,* 15
Energy
  energy processed in, 56–57
  stored in carbohydrates, 47
Envelope
  explanation of, **55**
Epithelial tissue
  in animals, 101, 102
Ethanol
  production of, 71
Ethical issue(s)
  human cloning as, 135
Eukarya, 20, *20*
Eukaryote(s)
  cell differentiation in, 99
  explanation of, 98
  reproduction in, 131
Eukaryotic cell(s)
  cellular respiration in, 70
  examination of, 54 *lab,* **87,** 98
  explanation of, 54
  organelle in, 55, 56
Exocytosis
  explanation of, *64,* **64**
External stimuli
  explanation of, 12, *12*

F

Facilitated diffusion
  explanation of, 63, *63*
Federal Bureau of Investigation (FBI), 95
Fermentation
  alcohol, 71
  explanation of, 70
  types of, 71, *71*
Fertilization
  explanation of, **117,** *117*
Fiber(s)
  explanation of, **100**
Fingerprinting
  explanation of, 95
Fission
  explanation of, *130,* **130**
Flagella
  explanation of, 53
Foldables, 10, 22, 29, 35, 37, 46, 55, 61, 70, 86, 99, 109, 118, 129, 141, 155, 164, 174, 181
Forensics
  explanation of, 95
Franklin, Rosalind, 171
Frogs
  reproduction in, 11, *11*
Fungi
  classification of, *20*
Fungus
  reproduction in, 129
Furrow
  explanation of, 92

G

G₁ interphase stage
  explanation of, 88
G₂ interphase stage
  explanation of, 88
Gene(s)
  environment and, 166, *166*
  explanation of, **160**, 170, 173
Genetic disorder(s)
  DNA mutations and, 176, *176*
Genetic variation
  explanation of, 124, *124*
  importance of, 136
Genetics. *See also* **Inheritance**
  DNA and, *170*, 170–176, *171, 172, 173*
  experimentation with, 178–179 lab
  explanation of, **149**, 160
  inferred from phenotypes, 161 lab
  pioneers in, 149, 157
  prediction of, 162, *162*
  symbols for, 161, *161*
Genus
  explanation of, 21
Glucose
  converted into adenosine triphosphate, 71
  explanation of, 57
Glycolysis
  explanation of, 69, *69*
Golgi apparatus
  explanation of, 57, *57*
Grizzly bear
  classification of, *21*, 22
Ground tissue
  in plants, 101, 102, *102*
Guanine
  in nucleotide, 171

H

Haploid cell(s)
  explanation of, **119**
Health-care field(s)
  microscope use in, 30
Heredity
  explanation of, **149**
  Mendel's experiments on, *150*, 150–155, *151, 152, 153, 154*
Hershey, Alfred, 157
Heterozygous
  explanation of, *161*, **161**
Homeostasis
  explanation of, 7, **13**, 45
  importance of, 13
  methods for, 13
Homozygous
  explanation of, *161*, **161**
Hooke, Robert, 27, 43
How It Works, 49
Human cloning
  ethical issues related to, 135
Human(s)
  blood types in, 165, *165*
  chromosomes in, 159, *159*, 173, 175
  healthful lifestyle choices by, 166
  organ systems in, 103, 104, *104*

Hybrid plant(s)
  cross-pollination of, 153, *153, 154*
  explanation of, **153**
Hybrid
  explanation of, **153**
Hydra
  reproduction in, 131, *131*, 132
Hydrangea(s)
  flower color in, *166*

I

Ichthyologist(s) 17
Incomplete dominance
  explanation of, *164*, **164**
Inheritance. *See also* **Genetics; Genotypes; Traits**
  blending, 149
  complex patterns of, *164*, 164–165, *165*
  Mendel's experiments on, *150*, 150–155, *151, 152, 153, 154*
  modeling, *162*, 162–163, *163*, 168
  polygenic, 165
Insertion mutation
  explanation of, 175, *175*
Internal stimuli
  explanation of, 12
Interphase
  explanation of, *86*, **86**, 87, *87*
  meiosis I and, 120
  organelles during, 89
  phases of, 88, *88*
Interpret Graphics, 16, 24, 31, 48, 58, 66, 73, 94, 105, 126, 137, 156, 167, 177

J

Junk DNA, 173

K

Keratin, 47
Key Concepts, 8, 18, 26, 42, 50, 60, 64, 68, 84, 96, 116, 128, 148, 158, 169
  Check, 14, 20, 22, 27, 29, 44, 47, 54, 57, 65, 70, 72, 86, 93, 99, 104, 121, 124, 129, 135, 136, 151, 155, 161, 163, 165, 170, 176
  Summary, 34, 76, 108, 140, 180
  Understand, 16, 24, 31, 36, 48, 58, 66, 73, 78, 94, 105, 110, 126, 137, 142, 156, 167, 177, 182
Kingdom(s)
  explanation of, **20**
  list of, 20
Komodo dragon, 97, *97*

L

Lab, 32–33, 74–75, 106–107, 138–139, 178–179. *See also* **Launch Lab; MiniLab; Skill Practice**
Lactic acid
  fermentation and, 71, *71*
Launch Lab, 9, 27, 43, 51, 61, 69, 85, 97, 117, 129, 149, 159, 170

Leeuwenhoek, Anton von, 27, *27*
Lesson Review, 16, 24, 31, 48, 58, 66, 94, 105, 126, 137, 156, 167, 177
Light microscope
  explanation of, **28**, 30
Light
  photosynthesis and, 71, 72, 74–75 lab
Linnaeus, Carolus, 20
Lipid(s)
  explanation of, 46, *46*, **47**
Living thing(s)
  characteristics of, 9, 15;
  classification of, *19*, 19–23, 19 lab, *21, 22, 23*
  energy use by, 14, *14*, 15
  growth and development of, 10–11, 15
  homeostasis in, 13, *13*
  organization of, 9, 15
  reproduction in, 11, 15
  responses to stimuli by, 12, 15

M

Macromolecule(s)
  in cells, 46, *46*
  explanation of, **45**
Map butterfly, *166*
Math Skills, 28, 31, 37, 65, 66, 79, 92, 94, 111, 123, 126, 143, 154, 183
Mating
  in spiders, 127
Meiosis
  characteristics of, *123*
  chromosome movement during, 138–139
  diploid cells and, 119, 119 lab, 122, *122*
  explanation of, **119**, 140
  haploid cells and, 119, 119 lab, 122, 123
  importance of, 122, *122*
  mitosis v., 123, *123*
  phases of, 119–121, *120, 121*
Meiosis I, 119, 120, *120*
Meiosis II, 119, 121, *121*
Mendel, Gregor
  background of, 149, 157
  conclusions reached by, 155, 160, 165
  experimental methods used by, *150*, 150–151, *151*, 164
  results of experiments by, *152*, 152–154, *153, 154*, 163
Meristem(s)
  explanation of, 100, *100*
  plant cloning and, 134
Metaphase
  meiosis I and, 120, *120*
  meiosis II and, 121, *121*
  mitosis and, 90, *90*
Microscope(s)
  advancements in, 44
  atomic force, 49
  compound, 7, 28, *28*
  development of, 27
  electron, 29, *29*, 30

MiniLab
    examining cells using, 43
    light, 28, 30
    use of, 30, 30 *lab*, 34
MiniLab, 12, 23, 30, 47, 54, 63, 93, 103, 133, 155, 161, 172. *See also* **Lab**
Mitochondria. *See* **Mitochondrion**
Mitochondrion
    explanation of, 56, *56*, 89, *89*
    reactions in, 70, *70*
Mitosis
    characteristics of, *123*
    chromosome movement during, 138–139
    explanation of, **89**, 93 *lab*, 119
    meiosis v., 123, *123*
    phases of, *90*, 90–91, *91*
Mitotic cell division
    explanation of, *131*, 131
Mitotic phase
    explanation of, 86, 89
    stages of, 86
Molecular analysis
    explanation of, 20
Molecule(s)
    water, 45, *45*
Monera, 20
mRNA
    explanation of, 174, 175
    made from DNA, 173, *173*
    protein made from, 174
Multicellular organism(s)
    cell differentiation in, 99, 99–100, *100*
    explanation of, **10**, 99
    growth and development of, 10, 11
    organs in, 102, *102*
    organ systems in, 103, 104, *104*
    tissues in, 101, *101*
Muscle tissue
    in animals, 101, 102
Mustard plant, 125, *125*
Mutation(s)
    explanation of, 136, **175**
    results of, 176, *176*
    types of, 175, *175*

N

National Human Genome Research Institute (NHGRI), 157
Nerve cell(s)
    shape of, 51, *51*
Nervous tissue
    in animals, 101, 102
Newt(s), 132
Nuclear envelope
    explanation of, 55
Nucleic acid(s)
    explanation of, *46*, **46**, 47 *lab*
Nucleolus
    explanation of, 55, 90
Nucleotide(s)
    changes in, 175
    DNA and, 171, *171*
    explanation of, *171*, **171**
    RNA and, 173, *173*
Nucleus
    explanation of, **55**

O

Offspring
    in asexual reproduction, 129, 132, 135, 136
    genetic makeup of, 122, 124
    physical characteristics in, 117 *lab*
    in sexual reproduction, 124, 129
Organ system(s)
    explanation of, **103**
    human, 103, 104, *104*
    plant, 103
Organ(s)
    explanation of, 102
Organelle(s), 54–57. *See also* **Cell organelles**
    explanation of, 86
    membrane-bound, 87
    replication of, 88, 89
Organism(s). *See also* **Living thing(s); Multicellular organism(s); Unicellular organism(s)**
    explanation of, 86
    classification of, *19*, 19–23, 19 *lab*, 21, 22, 23, 34
    energy use by, 14, *14*, 15
    explanation of, **9**, 34
    growth and development in, 10–11, 15
    homeostasis in, 13, 15
    multicellular, *99*, 99–104, *100*, *101*, *102*, *104*
    reproduction in, 11, 15
    responses to stimuli by, 12, 15
    unicellular, 98, *98*
Organization
    of living things, 97–104
    of systems, 97 *lab*
Osmosis
    explanation of, **62**
Ovarian cancer
    explanation of, *176*
Oxygen
    in organisms, 17

P

Paramecium
    explanation of, 13, *13*, 53
Passive transport
    explanation of, 61
Pea plant(s)
    genetic studies using, *150*, 150–154, *151*, *152*, *153*, 164
Pedigree(s)
    explanation of, 162, 163, *163*
Pedipalp(s), 127
Phenotype(s). *See also* **Inheritance**
    environment and, 166, *166*
    explanation of, **160**, *161*
    to infer genotypes, 161 *lab*
Phenylketonuria (PKU)
    explanation of, *176*
Phospholipid(s)
    in cell membranes, 52
    explanation of, 47

Photosynthesis
    chloroplasts in, 71
    explanation of, 57, 70, **71**
    importance of, 72, *72*
    light in, 71, 72, 74–75 *lab*
Planarian(s), 132, *132*
Plant cell(s)
    animal cells v., 59
    chloroplasts in, 57
    shape of, 51, *52*
Plant cloning
    explanation of, *134*, **134**
Plant(s)
    cells in, 100, *100*, 104
    hybrid, 153, *153*, 154
    organ systems in, 103
    photosynthesis and, 71–72
    true-breeding, 151, 152
Plantae, 20
Platnick, Norman, 127
Pollination
    in pea plants, 150, *150*
Polygenic inheritance
    explanation of, **165**
Potential
    explanation of, **132**
Prokaryote(s)
    cell division in, 130
    explanation of, 98
Prokaryotic cell(s)
    examination of, 54 *lab*
    explanation of, 54, 98
Prophase
    explanation of, 90, *90*
    meiosis I and, 120, *120*
    meiosis II and, 121, *121*
Protein(s)
    in cells, 47, 52, 55, 63
    explanation of, 46, *46*, **47**
    production of, 173, *173*
    transport, 63, *63*
Protista, 20
Punnett square(s)
    analysis of, 162, *162*
    explanation of, **162**
    use of, 163, 168

R

Reading Check, 9, 13, 28, 30, 47, 52, 61, 63, 69, 71, 88, 91, 92, 98, 101, 102, 103, 119, 123, 125, 130, 132, 152, 153, 160, 162, 166, 171, 172, 174
Recessive allele(s)
    explanation of, *161*, 161–163
Recessive trait(s)
    explanation of, **155**
Red blood cell(s)
    size and shape of, 51
Reflex
    explanation of, 12 *lab*
Regeneration
    explanation of, *132*, **132**
Replication
    explanation of, **172**

Reproduction
asexual, 129–136, *130, 131, 132, 133, 134, 135, 136,* 140
cell division as form of, 93
in organisms, 11
sexual, *117,* 117–125, *118, 120, 121, 122, 123, 124, 125,* 140

Respiration. *See* **Cellular respiration**

Review Vocabulary, 29, 44, 87, 124, 150. *See also* **Vocabulary**

Ribonucleic acid (RNA)
explanation of, 46, 47, 55, 173
protein production and, 173
types of, 174

Ribosome(s)
explanation of, 55, 56

rRNA
explanation of, 174

S

S interphase stage
explanation of, 88

Salt crystal(s)
dissolved in water, 45

Scanning Electron Microscope (SEM), 29, *29*

Schleiden, Matthias, 44

Schwann, Theodor, 44

Science & Society, 33, 95, 107, 139, 157, 179

Science Methods, 75

Science Use v. Common Use, 20, 55, 100, 134, 153. *See also* **Vocabulary**

Scientific name(s)
assignment of, 23 *lab*
explanation of, 21
use of, 22

Sea cucumber(s), 132

Sea star(s), 132

Sea urchin(s), 132

Selective breeding
explanation of, 125, *125*

Self-pollination
explanation of, 150, *150*
in true-breeding plants, 151

Sex cell(s)
explanation of, 117, 118
formation of, 119–122, *120, 121,* 125
variations in, 122

Sexual reproduction
advantages of, *124,* 124–125, *125*
disadvantages of, 125
explanation of, **117,** 129, 140
meiosis and, 123
in spiders, 127

Sheep
cloning in, 135, *135*

Shell(s)
examination of, 51 *lab*

Siamese cat(s), *166*

Sister chromatids
explanation of, 88, *88,* 89

Skill Practice, 25, 59, 67, 168. *See also* **Lab**

Species
explanation of, **21**

Sperm
explanation of, **117, 150**

Spider(s), 127

Spindle fiber(s)
explanation of, 90

Sponge(s), 132

Standardized Test Practice, 38–39, 80–81, 112–113, 144–145, 184–185

Stem cells
explanation of, **100**

Stimuli
explanation of, 12
external, 12, *12*
internal, 12
responses to, 12, 12 *lab*

Stolon(s), 133, *133*

Strawberry plant(s)
reproduction by, 133, *133*

Study Guide, 34–35, 76–37, 108–109, 140–141, 180–181

Substitution mutation
explanation of, 175, *175*

Sun
energy from, 14

Sutton, Walter, 157

System(s)
organization of, 97 *lab*

Systematics
explanation of, 20

T

Tadpole(s), 120, *120,* 121, *121,* 132
explanation of, 91, *91*
growth and development of, 10, *10, 11*

Theory
explanation of, **44**

Thymine
in nucleotide, 171

Tissue(s)
animal, 101, *101,* 102
explanation of, **101**
plant, 101, *101,* 102

Trait(s). *See also* **Inheritance**
codes used to determine, 170 *lab*
cross-pollination experiments and, *154*
dominant, 155
factors that control, *159,* 159–161, *160, 161*
mutations and, 176
prediction of, *162,* 162–163
uniqueness of, 149 *lab*
variations in, 159 *lab*

Transcription
explanation of, *173,* **173**

Translation
explanation of, *174,* **174**

Transmission Electron Microscope (TEM), 29, *29*

Transport membrane(s)
explanation of, *63,* 63, 67

Transport
active, 64, *64*
passive, 61

tRNA
explanation of, 174

U

Unicellular organism(s)
explanation of, **10,** 98, *98*
growth and development of, 10
homeostasis in, 13, *13*
reproduction in, *11*

Uracil
in RNA, 173

V

Vacuole(s)
explanation of, 57

Vascular tissue
in plants, 101, 102, *102*

Vegetative reproduction
explanation of, **133,** 133 *lab*

Vesicle(s)
explanation of, 57

Virchow, Rudolf, 44

Visual Check, 10, 19, 45, 62, 85, 86, 102, 124, 130, 133, 152, 162, 163, 175

Vitamin A
explanation of, 47

Vocabulary, 7, 8, 18, 26, 34, 41, 42, 50, 60, 68, 73, 76, 83, 84, 96, 108, 115, 116, 128, 140, 147, 148, 158, 169, 180. *See also* **Academic Vocabulary; Review Vocabulary; Science Use v. Common Use; Word Origin**
Use, 16, 24, 31, 35, 48, 58, 66, 73, 37, 94, 105, 109, 126, 137, 141, 156, 167, 177, 181

W

Water
in cells, 45, *45,* 62
diffusion of, 62

Watson, James, 157

Weather
life cycle and, 85

What do you think?, 7, 16, 24, 31, 41, 48, 58, 66, 73, 83, 94, 105, 115, 126, 137, 147, 156, 167, 177

Whittaker, Robert H., 20

Wilkins, Maurice, 171

Williams syndrome
explanation of, *176*

Word Origin, 13, 21, 30, 45, 53, 62, 71, 89, 101, 119, 130, 160, 175. *See also* **Vocabulary**

Writing In Science, 37, 79, 111, 183

Y

Yeast
reproduction in, 129 *lab,* 131

Z

Zebra fish, 132

Zygote
chromosomes and, 118
explanation of, *117,* **117**

Credits

Photo Credits

Front Cover Spine Photodisc/Getty Images; **Back Cover** Thinkstock/Getty Images; **Inside front,back cover** Thinkstock/Getty Images; **Connect Ed** (t)Richard Hutchings, (c)Getty Images, (b)Jupiter Images/Thinkstock/Alamy; **i** Thinkstock/Getty Images; **iv** Ransom Studios viii–ix The McGraw-Hill Companies; **ix** (b)Fancy Photography/Veer; **6–7** Vaughn Fleming/Garden Picture Library/Photolibrary; **8** Angela Wyant/Getty Images; **9** (t)Hutchings Photography/Digital Light Source, (b)Peter Cade/ Getty Images; **10** (l)BRUCE COLEMAN, INC./Alamy, (c)Mark Smith/Photo Researchers, (r)Gary Meszaros/Photo Researchers; **11** (l)Joe McDonald/CORBIS, (r)BRUCE COLEMAN INC./Alamy; **12** (l)John Kaprielian/Photo Researchers, (c)Gary Gaugler/The Medical File/The Medical File/Peter Arnold, Inc.; **13** (r)Hutchings Photography/Digital Light Source; **14** Michael Abbey/Visuals Unlimited; **15** (t)Digital Vision/PunchStock, (tc)Robert Clay/Alamy, (c)John Kaprielian/Photo Researchers, (c)Splinter Images/Alamy, (b)Lee Rentz, (bc)liquidlibrary/PictureQuest; **16** (t)Mark Smith/Photo Researchers, (c)Michael Abbey/Visuals Unlimited, (b)liquidlibrary/PictureQuest, (inset)D. Finn/American Museum of Natural History, (bkgd)Reinaldo Minillo/Getty Images; **17** Spencer Grant/Photolibrary; **18** (t)Hutchings Photography/Digital Light Source, (b)PhotoLink/Getty Images; **21** Shin Yoshino/Minden Pictures/Getty Images; **22** Mark Steinmetz; **24** Shin Yoshino/Minden Pictures/Getty Images; **25** (t)Frank Greenaway/Getty Images, (tc)Adam Jones/Danita Delimont Agency, (b)Jill Van Doren/Alamy, (bc)Mark Steinmetz; **26** Eye of Science/Photo Researchers; **27** (t)Hutchings Photography/Digital Light Source, (c) Dr Jeremy Burgess/Photo Researchers, (b)Imagestate/Photolibrary; **28** (l)Steve Gschmeissner/Photo Researchers, (r)JGI/Getty Images; **29** (l)Stephen Schauer/Getty Images, (cl)ISM/Phototake, (cr)A. Syred/Photo Researchers, (r)age fotostock/SuperStock; **30** (t) Dr Jeremy Burgess/Photo Researchers; Hutchings Photography/Digital Light Source; **31** (l)Steve Gschmeissner/Photo Researchers, (tc)JGI/Blend Images/Getty Images, (r)A. Syred/Photo Researchers, (b)ISM/Phototake, (bc)Stephen Schauer/Getty Images; **32** (l,r)Hutchings Photography/Digital Light Source; **33** Hutchings Photography/Digital Light Source; **34** (t)Joe McDonald/CORBIS, (b)age fotostock/SuperStock; **36** Creatas/PunchStock; **37** (l)The McGraw-Hill Companies, (r)Vaughn Fleming/Garden Picture Library/Photolibrary; **40** (c)Dr. Donald Fawcett/Visuals Unlimited/Getty Images; **40–41** Dr. Dennis Kunkel/Visuals Unlimited/Getty Images; **42** Tui De Roy/Minden Pictures; **43** (t)Hutchings Photography/Digital Light Source, (c)Bon Appetit/Alamy, (b)Omikron/Photo Researchers; **44** (tl)Tim Fitzharris/Minden Pictures/Getty Images, (tc)Biophoto Associates/Photo Researchers, (tr)James M. Bell/Photo Researchers, (bl,bc,br)Dr. Richard Kessel & Dr. Gene Shih/Visuals Unlimited/Getty Images; **45** FoodCollection/SuperStock; **46** Dr. Gopal Murti/Photo Researchers; **47** Ed Reschke/Peter Arnold, Inc.; **48** (t)James M. Bell/Photo Researchers, (b)Dr. Gopal Murti/Photo Researchers; **49** (l)VICTOR SHAHIN, PROF. DR. H.OBERLEITHNER, UNIVERSITY HOSPITAL OF MUENSTER/Photo Researchers, (r)NASA-JPL; **50** Eye of Science/Photo Researchers; **51** Hutchings Photography/Digital Light Source; **53** SPL/Photo Researchers; **54** Hutchings Photography/Digital Light Source; **55** Dr. Donald Fawcett/Visuals Unlimited/Getty Images; **56** (l)Dr. Donald Fawcett/Visuals Unlimited/Getty Images, (r)Dennis Kunkel / Phototake; **57** (l)Dr. R. Howard Berg/Visuals Unlimited/Getty Images, (r)Dennis Kunkel / Phototake; **58** Dr. R. Howard Berg/Visuals Unlimited/Getty Images; **59** (c)Macmillan/McGraw-Hill; (others)Hutchings Photography/Digital Light Source; **60** LIU JIN/AFP/Getty Images; **61** Macmillan/McGraw-Hill; **63** Hutchings Photography/Digital Light Source; **67** (c)Macmillan/McGraw-Hill; Colin Milkins/Photolibrary, (others)Hutchings Photography/Digital Light Source; **69** Hutchings Photography/Digital Light Source; **71** (t)Biology Media/Photo Researchers, (b)Andrew Syred/Photo Researchers; **72** Michael & Patricia Fogden/Minden Pictures; **73** Biology Media/Photo Researchers; **74** (tc,bc)Macmillan/McGraw-Hill, (others)Hutchings Photography/Digital Light Source; **76** Dr. Richard Kessel & Dr. Gene Shih/Visuals Unlimited/Getty Images; **79** Dr. Dennis Kunkel/Visuals Unlimited/Getty Images; **82–83** Robert Pickett/CORBIS; **84** Michael Abbey/Photo Researchers; **85** (t)Hutchings Photography/Digital Light Source, (b)Bill Brooks/Alamy; **87** (bl)Ed Reschke/Peter Arnold, Inc., (br)Biophoto Associates/Photo Researchers, (others)Dr. Richard Kessel & Dr. Gene Shih/Visuals Unlimited; **88** Don W. Fawcett/Photo Researchers; **89,90,91** (l)Ed Reschke/Peter Arnold, Inc.; **92** (l)P.M. Motta & D. Palermo/Photo Researchers, (r)Manfred Kage/Peter Arnold, Inc.; **93** Hutchings Photography/Digital Light Source; **94** (t)Biophoto Associates/Photo Researchers, (b)P.M. Motta & D. Palermo/Photo Researchers; **95** (l)Alex Wilson/Marshall University/AP Images, (r)Bananastock/Alamy; **96** (inset)Biosphoto/Lopez Georges/Peter Arnold, Inc., (bkgd)Biosphoto/Bringard Denis/Peter Arnold, Inc.; **97** (t)Hutchings Photography/Digital Light Source, (c)Steve Gschmeissner/Photo Researchers, (b)Cyril Ruoso/ JH Editorial/Minden Pictures; **98** (t)Wim van Egmond/Visuals Unlimited/Getty Images, (c)Jeff Vanuga/CORBIS, (b)Alfred Pasieka/Photo Researchers; **99** (l)Biophoto Associates/Photo Researchers; **101** (r)Andrew Syred/Photo Researchers; **102** (inset)Ed Reschke/Peter Arnold, Inc., (bkgd)Michael Drane/Alamy; **103** Hutchings Photography/Digital Light Source; **105** Wim van Egmond/Visuals Unlimited/Getty Images; **106,107** Hutchings Photography/Digital Light Source; **111** Robert Pickett/CORBIS; **114–115** Bill Coster/Getty Images; **116** Science Pictures Ltd./Photo Researchers; **117** Hutchings Photography/Digital Light Source; **120,121** (t)Ed Reschke/Peter Arnold, Inc.; **122** (t,c)Nature's Images/Photo Researchers, (b)Jeremy West/Getty Images; **124** (l)Rob Walls/Alamy, (r)Nigel Cattlin/Alamy; **125** (tl)Stockbyte/Getty Images, (tr)Craig Lovell/CORBIS, (c)Piotr & Irena Kolasa/Alamy, (bl)image100/SuperStock, (br)Wally Eberhart/Visuals Unlimited/Getty Images; **126** Bill Coster/Getty Images; **127** (bl)Greg Broussard, (bkgd)David Thompson/Photolibrary; **128** Dr. Brad Mogen/Visuals Unlimited; **129** Horizons Companies; **130** CNRI/Photo Researchers; **131** (l,c,r)Biophoto Associates/Photo Researchers; **133** (l)sciencephotos/Alamy, (t)Wally Eberhart/Visuals Unlimited, (r)DEA/G.CIGOLINI/Getty Images, (b)Jerome Wexler/Photo Researchers; **135** Roslin Institute; **136** Mark Steinmetz; **137** (t)Dr. Brad Mogen/Visuals Unlimited, (c)Roslin Institute, (b)Mark Steinmetz; **138** (t,c,b)Hutchings Photography/Digital Light Source; **139** Hutchings Photography/Digital Light Source; **140** (t)Science Pictures Ltd./Photo Researchers, (b)CNRI/Photo Researchers; **143** Bill Coster/Getty Images; **146–147** Tom and Pat Leesson; **148** Nigel Cattlin/Visuals Unlimited; **149** (tl)Geoff du Feu/Alamy, (tr)Ken Karp/McGraw-Hill Companies, (cl)Glow Images/Getty Images, (cr)Getty Images, (bl,br)The Mcgraw-Hill Companies; **150** (l,r)Wally Eberhart/Visuals Unlimited; **151** (t)Tom and Pat Leesson, (bl)DEA PICTURE LIBRARY/Photolibrary, (br)WILDLIFE/Peter Arnold, Inc.; **152,153** (purple flower)DEA PICTURE LIBRARY/Photolibrary, (white flower)WILDLIFE/Peter Arnold, Inc.; **157** (t)Pixtal/age Fotostock, (c)World History Archive/age Fotostock, (b)Tek Image/Photo Researchers, (bkgd)Jason Reed/Getty Images; **158** (l to r, t to b,2,5–10)Getty Images, (3)Punchstock, (4,11,12)CORBIS; **159** (t)Hutchings Photography/Digital Light Source, (b)Biophoto Associates/Photo Researchers; **161** (t,b)Martin Shields/Photo Researchers; **163** (l)Maya Barnes/The Image Works, (r)Bill Aron/PhotoEdit; **164** (l)Peter Smithers/CORBIS, (tr)Geoff Bryant/Photo Researchers, (c)Bill Ross/CORBIS, (bl)J. Schwanke/Alamy, (bc,br)Yann Arthus-Bertrand/CORBIS; **165** Chris Clinton/Getty Images; **166** (tl)Picture Net/Corbis, (tr)June Green/Alamy Images, (c)Carolyn A. McKeone/Photo Researchers, (bl,br)Hania Arensten-Berdys/www.gardensafari.net; **167** Chris Clinton/Getty Images; **168** (l,r)Kristina Yu/Exploratorium Store; **169** PHOTOTAKE Inc./Alamy; **170** Hutchings Photography/Digital Light Source; **178** (all)Hutchings Photography/Digital Light Source; **179** (tr)The McGraw-Hill Companies; (others)Hutchings Photography/Digital Light Source; **180** (cl)WILDLIFE/Peter Arnold, Inc., (purple flower)DEA PICTURE LIBRARY/Photolibrary; **183** Tom and Pat Leesson; **184** (bc)Gary W. Carter/CORBIS, (br)Alamy

Credits

Images, (others)Ted Kinsman/Photo Researchers; **185** (t)Alamy Images, (c)Frank Krahmer/zefa/CORBIS, (b)Geoff du Feu/Alamy; **186–187** William Osborn/Minden Pictures; **188** Florida Museum of Natural History photo by Eric Zamora ©2008; Hutchings Photography/Digital Light Source; **190** (l)B. A.E. Inc./Alamy, (r)The Natural History Museum/Alamy; **191** (l)Mark Steinmetz, (c)Tom Bean/CORBIS, (r)Dorling Kindersely/Getty Images; **195** Hutchings Photography/Digital Light Source; **196** Dorling Kindersely/Getty Images; **197** (t)Ryan McVay/Getty Images, (c)C Squared Studios/Getty Images, (b)Richard Broadwell/Alamy, C Squared Studios/Getty Images; Getty Images; Tim O'Hara/Corbis; **198** DLILLC/CORBIS; **199** Hutchings Photography/Digital Light Source; **200** (l)Jeffrey Greenberg/Photo Researchers, (c)David Hosking/Alamy, (r)Mark Jones/Photolibrary; **201** Chip Clark; **203** (l)Carey Alan & Sandy/Photolibrary, (c)Robert Shantz/Alamy, (r)Stan Osolinski/Photolibrary; **204** (l)Paul Sutherland/National Geographic/Getty Images, (t)NHPA/Photoshot, (r)Kay Nietfeld/dpa/CORBIS, (b)Mitsuhiko Imamori/Minden Pictures; **205** (l)ARCO/D. Usher/age Fotostock, (t)Photodisc/Getty Images, (r)Joe Blossom/NHPA/Photoshot, (b)Hutchings Photography/Digital Light Source; **206** (t)Mark Jones/Photolibrary, (tc)Chip Clark, (b)Don Mammoser/Bruce Coleman, Inc./Photoshot, (bc)Paul Sutherland/National Geographic/Getty Images; **207** (t)D. Parer & E. Parer-Cook, (b)B. Rosemary Grant/AP Images, (bkgd)Skan9/Getty Images; **208** Joseph Van Os/Getty Images; **209** Hutchings Photography/Digital Light Source; **214** T. Daeschler/The Academy of Natural Sciences/VIREO; **215** Joseph Van Os/Getty Images; **216** (t to b,5Hutchings Photography/Digital Light Source, (2–4)Macmillan/McGraw-Hill, (br)Mary Plage/Photolibrary, (cr)USGS; **218** Mark Steinmetz; **221** William Osborn/Minden Pictures; **SR-00–SR-01** (bkgd)Gallo Images - Neil Overy/Getty Images; **SR-02** Hutchings Photography/Digital Light Source; **SR-06** Michell D. Bridwell/PhotoEdit; **SR-07** (t)The McGraw-Hill Companies, (b)Dominic Oldershaw; **SR-08** StudiOhio; **SR-09** Timothy Fuller; **SR-10** Aaron Haupt; **SR-42** (c)NIBSC / Photo Researchers, Inc., (r)Science VU/Drs. D.T. John & T.B. Cole/Visuals Unlimited, Inc.; Stephen Durr; **SR-43** (t)Mark Steinmetz, (r)Andrew Syred/Science Photo Library/Photo Researchers, (br)Rich Brommer; **SR-44** (l)Lynn Keddie/Photolibrary, (tr)G.R. Roberts; David Fleetham/Visuals Unlimited/Getty Images; **SR-45** Gallo Images/CORBIS; **SR-46** Matt Meadows.

PERIODIC TABLE OF THE ELEMENTS

Legend:
- Element — Hydrogen
- Atomic number — 1
- Symbol — H
- Atomic mass — 1.01
- State of matter

States:
- 🎈 Gas
- 💧 Liquid
- ⬜ Solid
- ⊙ Synthetic

A column in the periodic table is called a **group**.
A row in the periodic table is called a **period**.

Group	1	2	3	4	5	6	7	8	9
1	Hydrogen 1 **H** 1.01								
2	Lithium 3 **Li** 6.94	Beryllium 4 **Be** 9.01							
3	Sodium 11 **Na** 22.99	Magnesium 12 **Mg** 24.31							
4	Potassium 19 **K** 39.10	Calcium 20 **Ca** 40.08	Scandium 21 **Sc** 44.96	Titanium 22 **Ti** 47.87	Vanadium 23 **V** 50.94	Chromium 24 **Cr** 52.00	Manganese 25 **Mn** 54.94	Iron 26 **Fe** 55.85	Cobalt 27 **Co** 58.93
5	Rubidium 37 **Rb** 85.47	Strontium 38 **Sr** 87.62	Yttrium 39 **Y** 88.91	Zirconium 40 **Zr** 91.22	Niobium 41 **Nb** 92.91	Molybdenum 42 **Mo** 95.96	Technetium 43 **Tc** (98)	Ruthenium 44 **Ru** 101.07	Rhodium 45 **Rh** 102.91
6	Cesium 55 **Cs** 132.91	Barium 56 **Ba** 137.33	Lanthanum 57 **La** 138.91	Hafnium 72 **Hf** 178.49	Tantalum 73 **Ta** 180.95	Tungsten 74 **W** 183.84	Rhenium 75 **Re** 186.21	Osmium 76 **Os** 190.23	Iridium 77 **Ir** 192.22
7	Francium 87 **Fr** (223)	Radium 88 **Ra** (226)	Actinium 89 **Ac** (227)	Rutherfordium 104 **Rf** (267)	Dubnium 105 **Db** (268)	Seaborgium 106 **Sg** (271)	Bohrium 107 **Bh** (272)	Hassium 108 **Hs** (270)	Meitnerium 109 **Mt** (276)

The number in parentheses is the mass number of the longest lived isotope for that element.

Lanthanide series

Cerium 58 **Ce** 140.12	Praseodymium 59 **Pr** 140.91	Neodymium 60 **Nd** 144.24	Promethium 61 **Pm** (145)	Samarium 62 **Sm** 150.36	Europium 63 **Eu** 151.96

Actinide series

Thorium 90 **Th** 232.04	Protactinium 91 **Pa** 231.04	Uranium 92 **U** 238.03	Neptunium 93 **Np** (237)	Plutonium 94 **Pu** (244)	Americium 95 **Am** (243)